T0320393

NONLINEAR TRENDING TIME SERIES

Theory and Practice

World Scientific Series on Econometrics and Statistics

Series Editors
Jiti Gao *(Monash University, Australia)*
Yongmiao Hong *(Cornell University, USA & Chinese Academy of Sciences, China)*
Shouyang Wang *(University of Chinese Academy of Sciences, China)*

Published

Vol. 2 *Nonlinear Trending Time Series: Theory and Practice*
by Li Chen (Xiamen University, China), Jiti Gao (Monash University, Australia) and Farshid Vahid (Monash University, Australia)

Vol. 1 *Spatial Econometrics: Spatial Autoregressive Models*
by Lung-Fei Lee (Ohio State University, USA and Shanghai University of Finance and Economics, China)

WORLD SCIENTIFIC SERIES ON ECONOMETRICS AND STATISTICS – VOL. 2

NONLINEAR TRENDING TIME SERIES

Theory and Practice

Li Chen

Xiamen University, China

Jiti Gao

Monash University, Australia

Farshid Vahid

Monash University, Australia

World Scientific

NEW JERSEY • LONDON • SINGAPORE • BEIJING • SHANGHAI • HONG KONG • TAIPEI • CHENNAI

Published by

World Scientific Publishing Co. Pte. Ltd.

5 Toh Tuck Link, Singapore 596224

USA office: 27 Warren Street, Suite 401-402, Hackensack, NJ 07601

UK office: 57 Shelton Street, Covent Garden, London WC2H 9HE

Library of Congress Cataloging-in-Publication Data

Names: Chen, Li (College teacher), author. | Gao, Jiti, 1962- author. | Vahid, Farshid, author.
Title: Nonlinear trending time series : theory and practice / Li Chen, Xiamen University, China, Jiti Gao, Monash University, Australia, Farshid Vahid, Monash University Australia.
Description: New Jersey : World Scientific, [2025] | Series: World scientific series on econometrics and statistics; vol. 2 | Includes bibliographical references and index.
Identifiers: LCCN 2024030253 | ISBN 9789811293344 (hardcover) | ISBN 9789811293351 (ebook) | ISBN 9789811293368 (ebook other)
Subjects: LCSH: Time-series analysis. | Time-series analysis--Data processing.
Classification: LCC QA280 .C44 2025 | DDC 519.5/5--dc23/eng/20240807
LC record available at https://lccn.loc.gov/2024030253

British Library Cataloguing-in-Publication Data

A catalogue record for this book is available from the British Library.

For any available supplementary material, please visit
https://www.worldscientific.com/worldscibooks/10.1142/13844#t=suppl

Desk Editors: Aanand Jayaraman/Pui Yee Lum

Typeset by Stallion Press
Email: enquiries@stallionpress.com

Printed in Singapore

Preface

Over the past four decades, there has been significant interest in understanding trending time series data. Initially, models and methods relied on parametric linear time series models with constant coefficients. However, these approaches are overly restrictive both in theory and application. Consequently, there has been a shift toward employing non- and semi-parametric methods and techniques to model time series data, accommodating potential nonlinearity and nonstationarity. Existing research books include Fan and Yao (2003), Gao (2007), and Teräsvirta *et al.* (2010).

The main differences and advantages of this book in relation to those available in the existing literature are (1) we retain standard linear models to capture linearity in variables but allow for nonlinearity in parameters; (2) the regressors involved in our models can be either stationary, locally stationary, or integrated nonstationary; (3) the coefficients are assumed to be unknown functions of time; and (4) the nonparametric kernel-based estimation method and procedure involved in each case are computationally tractable with closed-form expressions and user-friendly algorithms. Meanwhile, we also mention that there is a separate strand of the literature that assumes that regression coefficients are random processes and estimates them by Kalman filter or Bayesian algorithms.

This book focuses on several classes of trending time series models. The structure of this book is organized as follows: (a) Chapter 1 provides an introduction and overview of trending time series; (b) Chapter 2 starts with deterministic trending models and then discusses cointegration related to common stochastic trends; (c) Chapters 3 and 4 provide some detailed descriptions and discussions on time series regression models associated with weak trends and strong trends, respectively; (d) Chapter 5 offers explanations and discussions on methodologies for testing for common trends; and (e) Chapter 6 showcases the empirical relevance and applicability of the proposed models and methods from earlier chapters, employing climate data as a demonstrative example. The book also includes an appendix summarizing the key technical details involved in the establishment and proof of the main results of this book.

This book covers the main estimation and testing methods for trending time series with the necessary technical details. The applied aspect presented in Chapter 6 enables researchers and graduate students to keep abreast of developments in the field.

While the authors of this book have tried their best to reflect the work of many researchers in the field, some other closely related studies might still be inevitably omitted in this book. The authors therefore apologize for any omissions.

We would like to thank anyone who has encouraged and supported us to complete this book. Especially, we would like to thank our coauthors working in similar areas, Heather Anderson, Tingting Cheng, Chaohua Dong, Oliver Linton, Bin Peng, Peter CB Phillips, Wei Wei, Yayi Yan, and Deshui Yu for all insightful discussions and comments during the years.

Our special thanks go to our family members. Li Chen would like to thank his wife Xiu Xu and lovely daughter Ming Chen for their support.

Jiti Gao offers his thanks to his wife, Mrs Qun Jiang, and two lovely sons, Robert and Thomas, for their continuing support.

Farshid Vahid would also like to thank Heather Anderson for her support both personally and professionally.

Finally, Li Chen would also like to acknowledge the financial support of the National Natural Science Foundation of China (NSFC, Grant numbers 72103173, 72233002, 72033008). Thanks from Jiti Gao and Farshid Vahid also go to the Australian Research Council Discovery Grants Program (DP200102769) for its financial support.

Without such support and cooperation, it would not be possible for us to complete the writing of this book.

About the Authors

Li Chen is Associate Professor at the Wang Yanan Institute for Studies in Economics (WISE), Xiamen University in China. He received his PhD degree from Monash University in 2017. His research interests lie in nonstationary time series modeling and its applications in economics, finance, and climate change. Dr Chen has published several research papers in leading journals in econometrics and finance. He is also the principal investigator and participant of several projects granted by the National Natural Science Foundation of China (NSFC).

Jiti Gao is Professor and Donald Cochrane Chair of Business and Economics at the Department of Econometrics and Business Statistics, Monash University in Australia. He is an Elected Fellow of the Academy of Social Sciences in Australia (FASSA) since 2012. He also holds fellowships with the International Association of Applied Econometrics and the *Journal of Econometrics* and *Econometric Theory*.

Farshid Vahid is Professor at the Department of Econometrics and Business Statistics, Monash University in Australia. He is an Elected Fellow of both the Academy of Social Sciences in Australia (FASSA) (since 2014) and the International Association of Applied Econometrics since 2018.

Contents

Chapter 1

Introduction

> No one understands trends, but everyone sees them in the data.
>
> — Phillips (2005)

Time series models serve as indispensable tools for researchers across many disciplines, enabling the analysis and comprehension of intricate datasets. Within the realms of economics and finance, these models are extensively applied to study key economic variables, including stock prices, exchange rates, interest rates, and inflation. Moreover, in the realm of management science, researchers rely on these models to predict sales, production trends, and demand patterns. Climate change researchers also employ these models to investigate climate data, facilitating the comprehension and projection of temperature fluctuations, sea level variations, and the likelihood of the occurrence of natural disasters. Beyond these, time series models have wide-ranging applications in areas such as healthcare, environmental science, engineering, and other domains as well.

Recent years have witnessed significant progress in the development of econometric theories concerning time series models, leading to notable improvements in modeling capabilities. Researchers have made impressive strides in creating models that can effectively handle the complexities of nonlinear and nonstationary features present in the data. They have devised a wide array of theoretical models, encompassing linear and nonlinear specifications in stationary and nonstationary contexts, integrating parametric, nonparametric, and semiparametric structures. These models are extensively employed to capture the interconnections, correlations, and shared characteristics inherent within different time series datasets.

In particular, the establishment of rigorous estimation and inference methods has played a crucial role in reinforcing the effectiveness of time series models for many empirical research applications (see Fan and Gijbels, 1996; Gao, 2007; Teräsvirta *et al.*, 2010). By employing these methodologies, researchers have gained a more comprehensive understanding of the intricate dynamics present within the datasets, enabling them to derive more accurate and nuanced insights from the data. Furthermore, various mathematical theorems are developed to reveal the intricate statistical properties associated with these new models and methodologies. They not only provide researchers with a heightened level of confidence in their analyses but have also furnished them with a more reliable and resilient framework for conducting empirical studies with greater accuracy and reliability.

With the availability of large amounts of time series in the big data era, researchers are able to investigate a wide range of empirical questions related to issues of serial dependence, causal relationships, and reliable forecasts of economic and financial indicators. Though the advancements in time series modeling have resulted in numerous successful applications across different fields of study, some econometric theories become inapplicable when nonstationary trending time series are involved in the analysis. This challenge highlights the need for ongoing research and development in this area to establish innovative models and methods that are capable of dealing with trending time series. This book provides some insights into the econometric models and methods that can be used to analyze trending time series and explore either the long-run or the short-run relationships between them.

This chapter is organized as follows. We first provide an overview of trending time series analysis and give some empirical examples before we formally discuss the characteristics of deterministic and stochastic trends. Then, we introduce some statistical tests for the two kinds of trends and argue the difficulty in distinguishing between nonlinear deterministic trends and stochastic trends by a simulated example. Finally, we discuss the problems of using detrending methods, which are widely applied to convert nonstationary time series into stationary ones.

1.1 An Overview on Trending Time Series

In real-world scenarios, empirical trending time series are a common occurrence. For instance, the daily U.S. stock market index over the past two years, the sequence of the U.K.'s gross domestic product (GDP) at the end of each year since 1920, the annual population data of London since 1900, and the recorded global mean temperatures annually since 1850. When initially observing a plot of such nonstationary trending time series, the human eye naturally gravitates toward recognizing the prominent trend elements within them. These could manifest as a visible upward trajectory in the annual global mean temperature series or a sustained long-term decline in the stock market indices. Before delving into the modeling theories of trending time series, it is essential to address a fundamental question: "What exactly constitutes a trend?"

Harvey (1990) referred to the definition of "trend" in the *Concise Oxford Dictionary* that a trend in a time series is "a general direction and tendency" and "part of the series which, when extrapolated, gives the clearest indication of the future long-term movements in the series" in the view of prediction. He also defined that "a global trend may be represented by a deterministic function of time which holds at all points throughout the sample, while a local trend may change direction during the sample and it is the most recent direction which we want to extrapolate into the future." Phillips (2010) mentioned that the dictionary definition of the word "trend" originates from a nineteenth-century usage as "the general course of events or prevailing tendency". Trends are usually the dominant characteristic in much economic data. But nowadays, trends are still little understood and we do not have a precise and unambiguous definition for "trend" though this term is widely used in many fields. White and Granger (2011) held the opinion that there is no generally accepted definition of "trend" though economists can give plenty of examples of trends. But they suggest that "a trend should have a direction, be somewhat smooth and be monotonic throughout". Therefore, most econometricians hold the opinion that almost everyone knows what a trend looks like, but it is difficult for us to give a rigorous and precise definition of a trend in general.

Nevertheless, before we investigate trends and trending time series models in this book, we give a brief review of how the opposite, a stationary time series with no trend, is defined. A stationary time series, as its name suggests, usually fluctuates around a stable level in the long run, and its statistical properties remain unchanged over time. For example, an independent and identically distributed (*i.i.d.*) process or a white noise (WN) process is stationary since it exhibits the same statistical behavior regardless of when a particular piece of the time series sequence is observed. In most of the classical time series regression models, the involved covariates are usually limited to being either strictly stationary or weakly stationary. Specifically, a time series $\{X_t\}$ is said to be *strictly stationary* or *strongly stationary* if the k-dimensional joint probability distribution function (pdf) of $(X_1, \ldots, X_k)$ is the same as that of $(X_{t+1}, \ldots, X_{t+k})$ for all t, where k is an arbitrarily chosen integer. A time series $\{X_t\}$ is said to be weakly stationary or *covariance stationary*, or *second-order stationary*, if $\mathrm{E}(X_t^2) < \infty$, $\mathrm{E}(X_t) = \mu$ and $\mathrm{Cov}(X_t, X_{t+k}) = \gamma_k$, where both μ and γ_k do not depend on t for any integer k. In another perspective, if $\Delta^d x_t = \varepsilon_t$, where ε_t is a zero mean WN with finite variance and $\Delta = 1 - L$ is a differencing operator, then x_t is stationary when $d < 1/2$.

Figure 1.1(a) shows a time series simulated from an *i.i.d.* process with zero mean and unit variance. According to how it is generated, the sequence fluctuates around zero with constant variance and no serial correlation. Therefore, it is a stationary process. Figure 1.1(b) is an autoregressive process of order 1 (AR(1) process), where the autoregressive coefficient is 0.4 so that the marginal effect of a unit shock at time t to its subsequence value at time $t + j (j \geq 0)$ is 0.4^j, which decays exponentially fast when j gets large. The process is also stationary since the absolute value of the autoregressive coefficient is less than one.

The estimation methods and inference theories for stationary time series models are generally straightforward and simple to apply, unlike situations involving nonstationary time series in regression models. A random variable representing nonstationary time series, such as those exhibiting deterministic or stochastic trends, often experiences changes in their mean values or higher-order moments over time. Consequently, the associated estimation methods and asymptotic properties of nonstationary trending regression models become intricate and unconventional. As a result, specialized

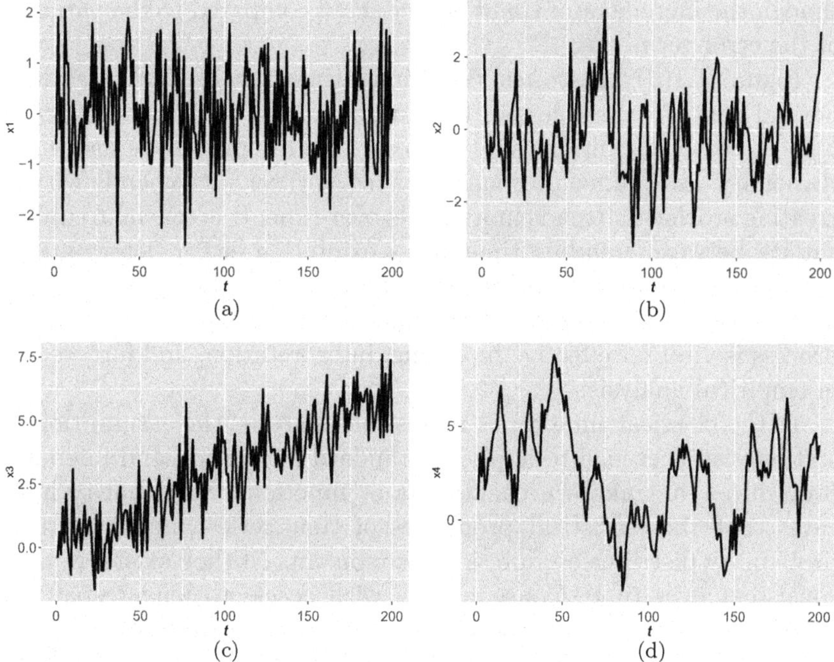

Figure 1.1. Simulated stationary and nonstationary time series: (a) An independent and identically distributed process $v_t \overset{i.i.d.}{\sim} \mathcal{N}(0, 1)$; (b) a stationary AR(1) process $x_t = 0.4x_{t-1} + v_t$; (c) a (linear) trend-stationary process $x_t = 0.03t + v_t$; (d) a random walk process $x_t = x_{t-1} + v_t$.

econometric theories must be developed to address nonstationary trending time series models.

In Figure 1.1(c), we observe one instance of a deterministic trending process. The process is simulated by $x_t = 0.03t + v_t$ for $t = 1, 2, \ldots, 200$, where the error term v_t follows independent and identically distributed (*i.i.d.*) standard normal distributions. This process is classified as nonstationary because its unconditional mean, $\mathrm{E}(x_t) = 0.03t$, evolves deterministically with the time index t. This indicates that the trend component in x_t is precisely determined for a given value of t. Notably, the trend component diverges when the sample size T tends to infinity. Hence, the linear deterministic trend component dominates the level of the simulated process in the figure though the increment of trend is very small compared to the variance of the error term.

Figure 1.1(d) illustrates the cumulative sum of *i.i.d.* standard normal random variables v_t, forming a random walk process $x_t = \sum_{s=1}^{t} v_t$ with an initial value of $x_0 = 0$. This process is also non-stationary, as its unconditional variance $\text{Var}(x_t) = t$ expands with t, creating stochastic trends in its levels over time. It is crucial to differentiate between trending time series exhibiting either deterministic or stochastic trends, as they carry distinct implications. As discussed in the subsequent section, comprehending the type of data-generating process is vital for effective modeling, interpretation, and forecasting in empirical analysis.

In the classical time series regression analysis, the assumption of stationarity is crucial to applying standard estimation and inference procedures for unknown coefficients or functions. Stationarity indicates that the statistical properties of time series, including their probability distribution functions or moments (if they exist), remain consistent over time. Consequently, with access to longer spans of sample series, we can leverage a more extensive dataset to accurately and effectively uncover the true statistical properties of the time series data. In other words, this expanded information from an extended dataset allows for a more comprehensive understanding of their underlying probability distributions, serial dependencies, and correlation relationships. Furthermore, a significant feature of stationary data is the decaying temporal dependence within a stationary time series sequence (e.g., x_t for $t = 1, 2, \ldots, T$). In particular, the short memory case that $\sum_{j=0}^{\infty} \gamma(j) < \infty$, where $\gamma(j) = \text{Cov}(x_t, x_{t-j})$ is the jth order autocovariance.

Stationarity implies asymptotic uncorrelatedness between x_t and x_{t+j} as the temporal distance between their time indices, $|j|$, approaches infinity. This property enables regression methods and theories for such stationary time series to resemble and align with those employed in classical regression analysis for *i.i.d.* data within the cross-sectional framework. However, it is necessary to adjust the corresponding econometric theories to accommodate time series characteristics, such as serial correlation and conditional heteroscedasticity.

At the same time, impulse responses and forecasting future values based on historical data reveal distinct characteristics for stationary and trending time series, respectively. In the case of a stationary time series, an unforeseen shock typically yields a short-lived effect that rapidly dissipates over time. Forecasts for this type of series at long horizons are relatively trivial as the stationary time series reverts to its constant mean, variance, and higher moments over time. Consequently, long-term forecasting for a stationary time series is always challenging because of limited information available in the distant future.

On the contrary, grasping the underlying trending patterns allows for a much more confident forecast of the future behavior of trending time series. To this end, identifying the dominant trend component in a nonstationary trending time series becomes crucial for enhancing forecasting precision, as forecasting trends essentially involves capturing the momentum of historical trending behaviors. However, it is important to recognize that trends may undergo changes over time, which complicates the accurate forecasting of trending time series.

In practical applications, empirical time series data frequently demonstrate trending behaviors that violate the stationarity assumption. For example, macroeconomic variables such as the GDP, aggregate personal income and consumption, and treasury security yield exhibit trend-like movements even after applying logarithmic transformations. Their time series plots depict the economy's long-term growth and changes. Similarly, financial data, including stock prices and market indices, may display prolonged upward or downward trends, reflecting investors' buying and selling activities and signaling bull and bear market phases.

Climatic and environmental time series data, such as historical global mean temperature records and greenhouse gas emissions, frequently reveal significant trends, notably over recent decades, demonstrating a combination of trend-like movements and random wandering behaviors documented in the literature. These trends hold critical implications for our understanding of climate change and environmental concerns, necessitating the utilization of specialized statistical techniques that can account for nonstationarity. In addition to trending behaviors, nonstationary time series can also exhibit seasonal patterns and periodic variations, further adding complexity to their analysis and modeling in practical applications.

Figure 1.2 shows four empirical trending time series. The top left panel presents the time series plot of the daily close price of the

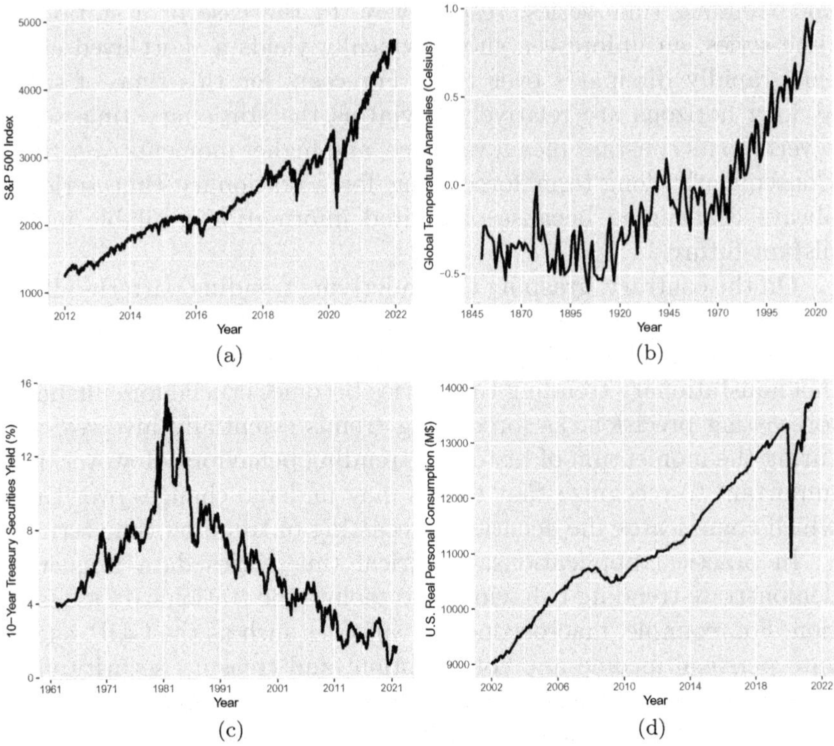

Figure 1.2. Four empirical trending time series data: (a) S&P 500 index; (b) global temperature anomalies; (c) ten-year treasury securities yield; (d) U.S. real personal consumption.

S&P 500 index from 1 January 2012 to 31 December 2021. The top right panel is the annual global mean temperature anomalies, which are computed as the differences of annual mean temperatures from the 1960–1990 average. The bottom left panel plots the daily market yield on U.S. treasury securities at the ten-year constant maturity from 1962 to 2021. The bottom right panel is the seasonally adjusted monthly real personal consumption expenditures from 2002 to 2021 with a unit of billions of chained 2012 U.S. dollars.

These four plots exhibit noticeable trending features, although they differ to some extent in patterns and magnitudes. Starting from 2012, both the S&P 500 index and the U.S. real personal consumption depict upward trends, albeit temporarily interrupted by significant events (Figures 1.2(a) and (d)). For instance, the emergence of the COVID-19 pandemic in early 2020 led to sharp declines in both series. However, in 2021, the series rebounded to their long-term trends, displaying a distinct dip in both plots. Therefore, it seems that the COVID-19 pandemic has caused a temporary shock to the S&P 500 index and the U.S. real personal consumption. Note that determining whether temporary shocks can result in temporary or permanent effects on a time series remains an enduring and intriguing question in trending time series analysis, and the answer in fact depends upon the underlying generating mechanism of the trending time series data.

Figure 1.2(b) shows that, between 1855 and 1920, the global temperature anomaly series exhibited relatively stable behavior, warranting its classification as a stationary time series during this period. From 1920 to 1944, the temperature series experienced a rapid increase of 0.5°C, followed by a mild cooling period until 1970. Subsequently, the global mean temperature has been escalating at an unprecedented rate, leading to the widely recognized phenomenon of global warming. This is usually attributed to excessive greenhouse gas emissions from human activities, and in the literature, there has been a debate on whether the temperature series in Figure 1.2(b) is trend-stationary or difference-stationary.

In contrast to the above three trending time series, in Figure 1.2(c), the ten-year treasury securities yield demonstrates an inverted V-shaped trend from 1962 to 2021 rather than a continuously rising or declining long-term trend. An evident turning point can be observed around 1981. Prior to 1981, the yield series had been ascending for 20 years since 1962, surpassing 15% in 1981. After 1981, the yield data displayed a prolonged downward trend for over 40 years. A large volume of papers discuss the reasons and the consequences of such turning in the trend of this important financial time series.

To summarize, Figure 1.2 highlights the dominant influence of trends in these nonstationary time series data, dictating their long-term evolution. Short-term shocks and intermediate deviations are relatively minor in comparison to the enduring trends, although they may cause temporary fluctuations contrary to the overall trend in localized periods. Moreover, the presence of turning points (breakpoints) is possible, although it often takes decades for the embedded trend to change direction.

Given the time-varying statistical properties of nonstationary time series, the violation of the stationarity assumption renders conventional estimation and inference methods inapplicable. Thus, the challenge of dealing with nonstationary time series data in classical regression models stems from the dynamic and fluctuating statistical properties exhibited by these series over time. To tackle this issue, one approach involves transforming nonstationary time series into stationary alternatives using techniques, such as differencing or detrending. However, this strategy can introduce other complications, including the potential loss of crucial information or the introduction of spurious relationships.

The primary objective of this book is to tackle the challenges associated with modeling trending time series, presenting readers with useful methods and theories from the literature. The overarching aim is to equip readers with a comprehensive toolkit designed for handling nonstationary trending time series within the context of regression analysis. This involves a review of recently established models, accompanied by an in-depth exploration of estimation and inference theories linked to these models. Consequently, readers are anticipated to access the essential methods and techniques necessary for effectively addressing nonstationary time series data. The following discussion outlines some key challenges encountered in the treatment of nonstationary trending time series.

First, the identification of the fundamental source behind trend behaviors, as emphasized by Phillips (2005, 2010), can often pose a significant challenge. Typically, trends evolve gradually over time, manifesting as low-frequency signals within the time series. These trends are propelled by underlying fundamental forces, including but not limited to economic factors, such as shifts in productivity, technological advancements, and systemic economic reforms. Similarly, fundamental climatic factors like solar energy influx, vegetation distribution, and the overall impact of greenhouse gases on global atmospheric conditions contribute to trend development. However, effectively pinpointing and confirming the key factor responsible for generating trend behavior within the data remains an intricate task. Without a comprehensive understanding of the fundamental drivers that give rise to these trends, the explanation and prediction of these trends become notably challenging.

Second, although the trending behaviors are evident as shown in the time series plots, the trend components are not explicitly observed and their generating process mechanisms are unknown (even unknowable) to us, as emphasized by Phillips (2001, 2003,

2005, 2010). In essence, the precise nature of these trends, whether deterministic, stochastic, or a blend of both, remains elusive, particularly when empirical researchers only have access to a finite sample of time series data. Meanwhile, their trends may also be confounded with other features of the time series, such as seasonality, serial correlation, and conditional heteroscedasticity in the error term of its data-generating process. Consequently, we may misunderstand the behaviors of economic variables as trends with different natures have entirely different empirical interpretations. Moreover, detrending operations, such as filtering or differencing, can pose significant risks if the data-generating mechanism is incorrectly specified.

Third, as either deterministic or stochastic trends are embedded in the time series, the probability distributions and characteristics of the nonstationary time series variables change accordingly over time. At the same time, as trends are more elusive to model empirically, reliable forecasts are much harder to estimate if the trend mechanisms are poorly specified and captured (Phillips, 2005). Furthermore, our knowledge of trends may be highly incomplete when only a limited time span of data is available. In other words, the trending component can be endogenous and dependent on the sample size of the data. Put simply, trends, as one has observed within the limited span of time series observations, may turn out to be part of a more extended periodic pattern rather than a linear trend when a longer span of data further unfolds. Alternatively, it could also stem from a segment of an autoregressive process with persistence close to but less than unity. However, as more time series data become accessible to the empirical analyst, the previously apparent trending pattern may tend to dissipate. We provide two examples in Figure 1.3 to illustrate this phenomenon.

The top panel of Figure (1.3) plots a simulated time series generated by $y_{1t} = 2.5\sin(4.35\pi t/n + \pi/4) + v_t$, for $v_t \overset{i.i.d.}{\sim} \mathcal{N}(0,1)$, $t = 1, 2, \ldots, n$. Suppose one can only access the initial quarter of the data (depicted as the solid line), and the real data-generating mechanism is unknown to the analyst. It is plausible that one might erroneously deduce the existence of a robust downward trend within the data. However, when a more extensive data span becomes accessible, it is revealed that the actual data-generating mechanism follows a periodic sine function of time with error disturbances. In this context, the perceived trend is, in fact, a part of the periodic cycles

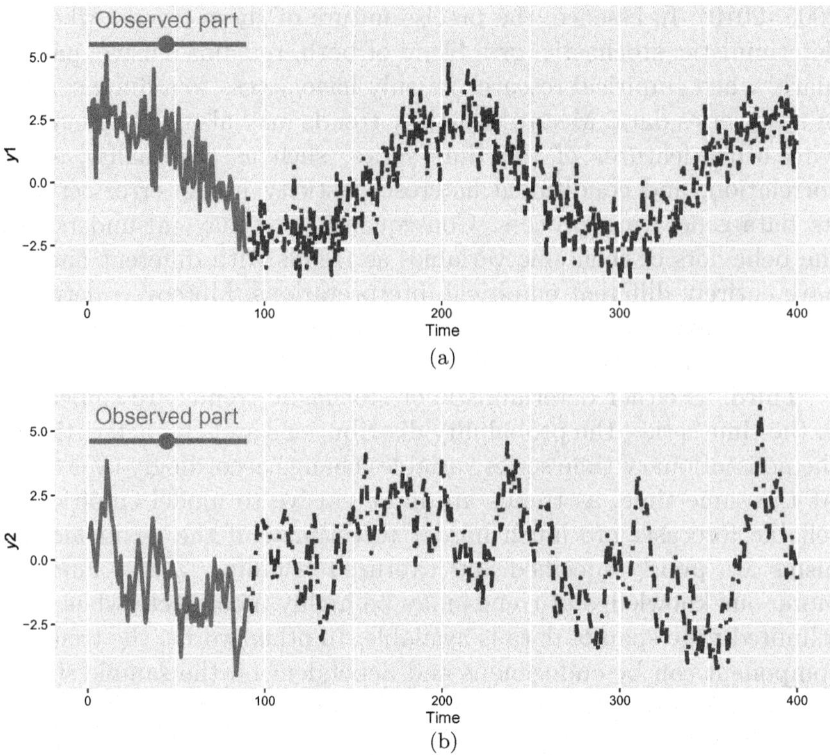

Figure 1.3. Partial observation of (a) $y_{1t} = 2.5\sin(4.35\pi t/n + \pi/4) + v_t$; (b) $y_{2t} = 0.9y_{2,t-1}+v_t$, for $v_t \overset{i.i.d.}{\sim} \mathcal{N}(0,1)$, $t = 1, 2, \ldots, n$. The solid line represents the observed part of the data, while the dashed line represents the unobserved part of the data.

inherent within the data. Consequently, we have to be aware that our understanding of the nature and characteristics of the trend component in the empirical data may not remain unchanged with respect to the length of the available sample.

The bottom panel of Figure (1.3) plots a simulated time series by $y_{2t} = 0.9y_{2,t-1} + v_t$, for $v_t \overset{i.i.d.}{\sim} \mathcal{N}(0,1)$, for $t = 1, 2, \ldots, n$. Again, if only the initial quarter of data is available to the empirical researchers who have no prior information on the real data-generating mechanism, they may falsely infer that a significant declining trend is present. However, this simulated time series is stationary, with neither deterministic nor stochastic trends. The misleading perception of a declining trend emerges from the partial observation of the data, where a substantial negative shock takes time to dissipate. It is common to observe that an autoregressive process structure characterized by relatively high persistence may exhibit trend-like behaviors during a short period of time. Therefore, making premature judgments about the presence and features of trends from partial observations of time series necessitates substantial caution in practical applications.

Lastly, the association between multiple trending time series often presents complex challenges. As stated by Phillips (1998), empirical trending time series are more susceptible to exhibiting spurious correlations, primarily because their respective trending components may showcase seemingly similar behaviors, even though they stem from distinct and independent data-generating processes. This phenomenon often leads to the potential pitfalls of drawing inaccurate empirical conclusions when employing standard estimation and inference methods based on trending time series data. Therefore, it becomes essential to devise and implement robust econometric methodologies capable of discerning genuine long-run stable relationships between various trending time series, particularly in cases where the specific nature of these trends remains somewhat elusive or ambiguous. One effective strategy to address this issue involves the rigorous testing of whether the time series in question share common underlying trends over the extended time horizon. For example, by examining the presence of cointegration (the existence of common stochastic trends), researchers can better assess the true nature of the relationships between these trending time series, thereby enhancing the reliability and validity of their empirical analyses and conclusions. In subsequent sections of this book, we demonstrate that trends are multifaceted and not confined to stochastic trends alone. Therefore, we introduce concepts that accommodate the presence of common complex trends among these trending time series. This approach represents a more comprehensive and general perspective compared to the conventional understanding of cointegration.

In this book, a typical nonstationary trending time series could be decomposed and represented in an additive form as

$$\text{Trending time series} = \text{Trend} + \text{Shocks}, \tag{1.1}$$

where we may or may not know the exact form of the trend term, but it is the source of nonstationarity. Apart from that, shocks are either strongly or weakly stationary and fluctuate around the trend term.

For identification purposes, the mean of shocks is usually set as zero so that the trending time series and the trend term share the same mean. Note that multiplicative (exponential) trending forms, such as the S&P 500 index in Figure 1.2, could be transformed into additive alternatives as above by taking logarithms. In the literature, cyclical components that depict medium-term periodic signals may also be contained in the trending time series structure:

$$\text{Trending time series} = \text{Trend} + \text{Cycle} + \text{Shocks}. \tag{1.2}$$

However, we only focus on the form of equation (1.1) and do not bother with cycle components in this book.

The specialized econometric models devised for the analysis of trending time series are commonly known as "trending time series models." These models serve a crucial role in unveiling and elucidating the underlying trending behaviors within time series data while also shedding light on the intricate relationships that exist between these trends. By formulating and deploying effective trending time series models, researchers gain valuable tools to delve into historical trends, gain insights into the current dynamics, and make informed predictions about the future, particularly in the context of nonstationary trending data across various domains and disciplines.

For example, the application of these models enables us to not only identify and comprehend the pressing crisis of climate change through the scrutiny of historical data pertaining to global temperatures and greenhouse gas emissions but also grasp the immediate impacts of the COVID-19 pandemic on a country by meticulously analyzing daily data concerning infections and fatalities. Moreover, these models empower us to make well-informed investment decisions by accurately predicting future trends in stock prices and other financial assets, thereby enabling us to capitalize on favorable market conditions and opportunities.

1.2 Deterministic and Stochastic Trends

The challenges of trending time series econometrics have been extensively discussed and explored by Phillips (2001, 2003, 2005, 2010); Rao (2010); White and Granger (2011) and Cochrane (2012). These studies discuss various issues, which include defining different types of trends, isolating components in real data, accounting for structural breaks, and understanding the impact of endogeneity on trend analysis. As said by Phillips (2005), trends are full of mysteries that "*no one understands trends, but everyone sees them in the data*", which is the quote at the beginning of this chapter, and it is regarded as one of the laws and limits of modern econometrics.

In practice, adequately modeling trends can be extremely challenging. To ensure accurate modeling of their behavior and to make reliable predictions about future trends using time series models, it is crucial to grasp the basic nature of trends. Broadly speaking, trends can be categorized as either deterministic or stochastic, contingent upon the underlying mechanisms that generate the data.

A deterministic trend is one that always maps a given time point to a deterministic value as the level of trend. In other words, there is a clear and consistent pattern or direction to the trend. On the other hand, a stochastic trend is one in which the trend value is random at a given time point, meaning that it cannot be predicted with certainty. Distinguishing between these types of trends is crucial for selecting appropriate modeling techniques and making accurate predictions. Deterministic trends can often be modeled using simpler techniques, while stochastic trends require more complex models that account for the random nature of the trend.

Specifically, a deterministic trend refers to a systematic and predictable movement in the mean value of a time series over time without any randomness or uncertainty. It is apparent that polynomial forms of time trends with constant coefficients are deterministic. For example, a linear trend such as $g(t) = 1 + 2t$ or a quadratic trend such as $g(t) = 1 + 2t + t^2$, where t denotes time. In the literature, the *trend-stationary (TS) process* is composed of a linear deterministic time trend and a stationary process. Specifically, this kind of process is usually written as

$$x_t = a + bt + v_t, \tag{1.3}$$

for $t = 1, 2, \ldots, T$, where v_t is a sequence of stationary disturbances with mean zero variance $\sigma_v^2 (0 < \sigma_v^2 < \infty)$. Note that the expectation of x_t is $E(x_t) = a + bt$, and the unconditional variance of x_t is $\mathrm{Var}(x_t) = \sigma_v^2$, which is a constant that does not change with t. It is named *trend-stationary* because one can obtain a stationary time series by removing the component of a linear time trend.

The marginal effect of the innovation v_t on $x_{t+j}(j > 0)$, the subsequent values of the trend-stationary process, dies out quickly when serial correlation in v_t is weak. Therefore, the shocks v_t only generate temporary effects on the time series of x_t. In other words, once deviates from the linear trend, the trending time series tends to revert to the linear trend promptly.

As suggested by Harvey (1997), when the time series covers a long span of time, it is too restrictive to assume that the level and the slope of the trend components are fixed as constants for the whole sample period under investigation. That is, the trend component can hardly be specified as a simple linear function of time unless the length of the time series is relatively short. Hence, we adopt a nonlinear time trend specification as

$$x_t = g(t;\theta) + v_t, \tag{1.4}$$

for $t = 1, 2, \ldots, T$, where v_t is a properly defined stationary process and $g(t;\theta)$ is a nonlinear parametric function of time t, which is prespecified according to economic theory or primary data analysis. For example, a piece-wise linear function of time that accounts for breaks or a logistic function that captures smooth changes over time.

Given that trends are likely to be complex components in the empirical time series, any subjectively selected parametric forms may be incorrectly specified for them. Therefore, in a more general manner, we could allow the nonlinear function $g(\cdot)$ to be a nonparametric function without specifying certain functional forms. That is,

$$x_t = g(\tau_t) + v_t, \tag{1.5}$$

for $t = 1, 2, \ldots, T$, where $\tau_t = t/T$. The nonparametric term $g(t)$ is expected to accommodate the nonlinear and nonstationary characteristics of the data. Though their functional forms are not exactly defined, they are still deterministic trends.

On the other hand, a stochastic trend refers to a time series that exhibits a random, persistent, and long-term movement in its mean value over time. Stochastic trends are often encountered in macroeconomic and financial time series, where long-term economic trends are affected by random shocks, such as changes in technology, policy interventions, or shifts in consumer preferences. Unlike a deterministic trend, a stochastic trend has an element of randomness or uncertainty that makes it difficult to forecast or predict the future behavior of the series. In other words, random shocks or innovations drive the time series with stochastic trends, and they generate permanent effects on the series over time. As a result, time series with stochastic trends become non-stationary, and the statistical properties of the series change over time. Therefore, standard regression techniques may not be appropriate for modeling and analyzing the data, and understanding and modeling stochastic trends is important for predicting and analyzing the behavior of economic variables over the long term.

In discriminating stationary and nonstationary time series, $I(0)$ and $I(1)$ are two frequently used terminologies in the literature. In the works of Breitung (2002); Davidson (2002); Müller (2008); Stock (1994), a time series $\{y_t\}_{t=1}^T$ is $I(0)$ if and only if there exists some $\sigma > 0$ such that for $T \to \infty$ and any $r \in [0,1]$, $T^{-1/2}\sigma^{-1}\sum_{t=1}^{[Tr]} y_t \Rightarrow W(r)$; meanwhile, time series $\{y_t\}_{t=1}^T$ is $I(1)$ if and only if there exists some $\sigma > 0$ such that $T^{-1/2}\sigma^{-1}y_{[Tr]} \Rightarrow W(r)$, for $T \to \infty$ and $r \in [0,1]$, where $[.]$ is the largest smaller integer function, $W(\cdot)$ is a standard Wiener process, and $\Rightarrow$ indicates weak convergence. The two definitions show that an $I(0)$ process is a possibly heterogeneous process that satisfies a functional central limit theorem (FCLT), and an $I(1)$ process is any process whose increments are $I(0)$. In general, y_t is an integrated process of order d ($I(\mathrm{d})$ process) when $\Delta^d y_t$ is $I(0)$ for some $d \geq 0$ in which Δ is the differencing operator.

In the literature, a widely entertained form of trending time series is formulated as

$$x_t = \delta + x_{t-1} + v_t, \tag{1.6}$$

for $t = 1, 2, \ldots, T$, where δ is called the drift term and v_t is again a stationary process with zero mean and variance $\sigma_v^2 (0 < \sigma_v^2 < \infty)$. It is the well-known random walk process with $(\delta \neq 0)$ or without $(\delta = 0)$ a drift. It is also a simple example of a unit root process as unity is the root of $1 - z$, the characteristic function of this autoregressive process. As stationarity could be easily achieved by taking the first-order difference of the time series that $\Delta x_t = x_t - x_{t-1} = \delta + v_t$, it is an $I(1)$ process and also named a *difference-stationary process.*

The data generating process of equation (1.6) indicates that x_t consists of a constant, the entire value of x_{t-1} without any loss, and the innovation at time t. By iterating the expression of $x_{t-1}, x_{t-2}, \ldots,$ we obtain an alternative expression of this process as

$$x_t = x_0 + \delta t + \sum_{s=1}^{t} v_s, \tag{1.7}$$

where x_0 is the initial value of this time series sequence. According to this, we find that the unconditional expectation of x_t is $E(x_t) = x_0 + \delta t$, and the unconditional variance of x_t is $\mathrm{Var}(x_t) = t\sigma_v^2$. Since the variance expands with t, it generates a stochastic trend. This new equation indicates that by accumulation, a non-zero drift term δ would form a linear deterministic time trend $x_0 + \delta t$ in the long run. Meanwhile, all the innovations till time t are accumulated and preserved in x_t. In other words, $v_{t-k}(0 \leq k < t)$ has permanent impacts on x_t no matter how large the distance k is. Hence, there are sharp differences between the economic interpretations of the two kinds of data-generating processes in equations (1.3) and (1.6).

A combined form of deterministic and stochastic trends may also appear in practice. For example, a state space model by Busetti and Harvey (2008) based on the local level model by Harvey (1990):

$$x_t = \mu_t + v_t, \tag{1.8}$$

$$\mu_t = \mu_{t-1} + \beta_t + \eta_t, \tag{1.9}$$

$$\beta_t = \beta_{t-1} + \xi_t, \tag{1.10}$$

for $t = 1, 2, \ldots, T$, with v_t, η_t, ξ_t following normal distributions with zero mean and variances σ_v^2, σ_η^2, and σ_ξ^2, respectively. This model nests different trending time series expressions over different choices of σ_η^2 and σ_ξ^2. That is, both, one, or none are equal to zero, which will lead to the different studied models. For example, $\sigma_\eta^2 = 0$ and $\sigma_\xi^2 = 0$ represent deterministic trend in x_t. When $\sigma_\eta^2 > 0$ and $\sigma_\xi^2 = 0$, it means x_t has a stochastic level and deterministic slope. When $\sigma_\eta^2 = 0$ and $\sigma_\xi^2 > 0$, it means x_t has a deterministic level and stochastic slope. When $\sigma_\eta^2 > 0$ and $\sigma_\xi^2 > 0$, both the level and slope of x_t are stochastic.

1.3 Testing for Deterministic and Stochastic Trends

Econometricians have made long-term efforts to develop useful tools that can be used to distinguish between stationary and different types of trending time series. Unit root processes are commonly encountered in economics and finance, where they are used to model nonstationary time series, such as stock prices, exchange rates, and inflation rates. The term "unit root" refers to the fact that the root of the characteristic equation of the process is equal to one. This means that the process has a unit root or a root of order one. A unit root process exhibits stochastic trending behaviors as the variance of the series is not constant over time.

For a sequence of empirical time series, the exact data-generating process is usually unknown. To distinguish between a unit root process and a (trend-) stationary process, econometricians have established various unit root tests since the 1980s. The *Augmented Dickey–Fuller (ADF) unit root test* and the *Phillips–Perron (PP) unit root test* are the most widely used statistical tests; see Fuller (1976), Dickey and Fuller (1979), and Phillips and Perron (1988). The null and the alternative hypotheses in these tests are established as

$$\mathbb{H}_0: \quad x_t \text{ has a unit root,} \tag{1.11}$$

$$\mathbb{H}_1: \quad x_t \text{ is (trend) stationary,} \tag{1.12}$$

where x_t represents the empirical time series data to be tested. When rejecting the null hypothesis, we obtain statistical evidence against a unit root. In other words, the underlying process is stationary. Note that a deterministic time trend can be included in the test equation. Therefore, the alternative hypothesis is allowed to be a linear trend-stationary process as many of the unit root tests (for example, the ADF test) only allow for a linear deterministic time trend.

Testing for unit roots in the empirical time series data is a necessary step before applying the time series models. Economists are aware that unlike in the stationary case, the estimation and inference of the coefficients are greatly different when the time series contain unit roots. In the work of Nelson and Plosser (1982), 14 U.S. macroeconomic time series are examined using the ADF test, and for most of the time series, they failed to reject the null hypothesis of unit root.

This paper has been extensively discussed and cited, and similar results are also found by Said and Dickey (1984) and Perron (1988), where the error terms in the test equation are allowed to be serially correlated.

Given that unit root tests do not suffer from severe size distortion, the probability of making the *Type I error* could be controlled by the pre-selected significance level. That is, under 5% significance level, the probability of incorrectly rejecting the null hypothesis $\mathbb{H}_0$ is less than 5%. Hence, it means that rejecting the null hypothesis is usually statistically reliable to confirm a (trend-) stationary process. Nevertheless, size distortions may still happen in some rare cases where we over-reject the null hypothesis. For example, we may over-reject the null hypothesis of a unit root when the time series is generated by $y_t = y_{t-1} + u_t$, where $u_t = \varepsilon_t + \theta\varepsilon_{t-1}$ in which ε_t is a WN process and θ is large and negative. The reason is that the sample path of y_t may behave more like a stationary process rather than a random walk (see Schwert, 1989).

However, the probability of making the *Type II error* may be substantially large if the unit root test has power problems. In other words, a low-power test favors the unit root null so that it is rarely rejected even though the null hypothesis is false while the alternative hypothesis is true. For example, the ADF tests are found to have low power against the alternative hypothesis of the autoregressive process with roots near unity in DeJong *et al.* (1992). The power is less than 30% when the autoregressive coefficient is over 0.9 but smaller than 1 when the sample size is around 100. Elliott *et al.* (1996) proposed the *ADF–GLS unit root test* to improve power when an unknown mean or trend is present. For the empirical side, Rudebusch (1993) found that the existence of a unit root is quite uncertain in the U.S. real GNP. Diebold and Rudebusch (1991) found that the ADF test also has low power against the alternative of fractionally integrated time series. Therefore, in such unit root tests of (1.11) and (1.12), failing to reject the null hypothesis does not necessarily imply the existence of unit root in the time series due to possible power problems. See the survey by Phillips and Xiao (1998) for more details on unit root tests.

The solution to this problem is to swap the null and the alternative hypothesis. Namely, we test a stationary null hypothesis against a unit root alternative. In this way, the probability of falsely accepting the alternative hypothesis of unit root could be controlled by the significance level we set, given that the test does not suffer from severe size distortion. By doing so, in the work of Kwiatkowski *et al.* (1992), the authors proposed the Kwiatkowski–Phillips–Schmidt–Shin (KPSS) unit-root test, in which the unit-root process is used as the alternative hypothesis. Thus, rejecting the null hypothesis strongly supports the existence of a unit root in the time series. In the work of DeJong and Whiteman (1991), the authors employed the Bayesian methods and found that only two of the Nelson–Plosser series contain unit roots. Even so, if we fail to reject both null hypotheses in the ADF-type and KPSS tests, we are still uncertain about the existence of unit roots in the data.

As discussed previously, the unit root conclusion for an empirical time series may be questionable due to the power problems of the unit-root tests. In practice, if the data exhibits an upward or downward trend, a linear function of time is usually included in the test equation of the ADF-type tests. Specifically, their test equation is usually formulated as

$$\Delta x_t = \underbrace{a + bt}_{\text{linear trend}} + \rho x_{t-1} + \alpha_1 \Delta x_{t-1} + \cdots + \alpha_p \Delta x_{t-p} + e_t, \qquad (1.13)$$

where $a + bt$ is a deterministic component used to capture the linear trend in the data. We reject the null hypothesis of unit root when the estimated value of $\widehat{\rho}$ is smaller than the critical value. The estimation of ρ and then the test result may be largely affected without including the deterministic trend component, which actually exists in the data.

However, Harvey (1997) suggested that for the empirical data, the deterministic trend is not necessarily linear. Ouliaris *et al.* (1989) developed a unit root test in a time series model that explicitly allows for polynomial trends and drift in the data generation process. They confirmed the importance of carefully modeling the deterministic component of a time series when testing for a unit root and found that some of the series can be modeled as stationary processes around a polynomial trend rather than a unit root process. Bierens (1997) used Chebyshev polynomials to replace the linear deterministic trend terms in the test equation so that the test allows for nonlinear deterministic trends. By doing so, some empirical time series, which were originally viewed as unit root processes, are tested and categorized as stationary variations around nonlinear functions of time.

Breaks are another type of nonlinearity in the deterministic trends. When allowing for a single break in the linear time trends, the test equation is formulated as

$$x_t = a + bt + \gamma DT_t(\tau_0) + u_t, \tag{1.14}$$

$$u_t = \rho u_{t-1} + \xi_t, \tag{1.15}$$

where $DT_t(\tau_0) = 1(t > \lfloor \tau_0 T \rfloor)(t - \lfloor \tau_0 T \rfloor)$, and the break point is given by $\lfloor \tau_0 T \rfloor$ in which $\tau_0 \in (0,1)$ is the break fraction and γ is the break magnitude. Our objective is to test $\mathbb{H}_0 : \rho = 1$ against the alternative that $\mathbb{H}_1 : \rho < 1$. Perron (1989) established a unit-root test in which a single trend break is allowed in both the null and the alternative hypotheses, where the location of the breakpoint is exogenously determined. In the unit-root test developed by Zivot and Andrews (1992), breakpoints were assumed to be endogenous to the data so that it is harder to reject the null hypothesis of the unit root as the test chooses the breakpoint location through the least favorable view of a unit root. Some recent developments of unit-root tests with breaks include the works of Harris *et al.* (2009), Cavaliere *et al.* (2011), Harvey *et al.* (2012), and Harvey *et al.* (2014), among others. Based on the unit-root test that allows for nonlinear deterministic trends, we should reject the null hypothesis of unit root for some of the time series in the Nelson–Plosser dataset, which, however, are not rejected by the usual tests with simple linear trends.

Instead of testing for stochastic trends, tests for the existence of deterministic time trends are also developed in the literature. The key issue is the consistent estimation of the long-run variance of the error term as it affects the statistical significance of the deterministic trend component. Elliott (2020) proposed new tests for the existence of a deterministic linear time trend in a trending time series, where there exists a potentially strong serial correlation in the errors. These tests can be applied as a pre-test for the necessity of including a time trend in a regression. Specifically, the paper considers the null hypothesis of $\mathbb{H}_0 : b = 0$ against the alternative $\mathbb{H}_1 : b > 0$ based on the regression model

$$x_t = a + bt + u_t, \tag{1.16}$$

$$u_t = \rho u_{t-1} + \xi_t, \tag{1.17}$$

for $t = 1, 2, \ldots, T$, where x_t is the observed sequence, $\{a, b, \rho, \xi_1\}$ and the variance of ξ_t are not observed. The autoregressive coefficient ρ

could be one, representing a mixture of deterministic and stochastic trends in x_t, or far below one, indicating a trend-stationary process for x_t. But equations (1.16) and (1.17) do not allow for nonlinear deterministic trends.

Estimation and testing theories on complex forms of deterministic time trends other than just linear, piece-wise linear, or quadratic ones are also established. For example, nonlinear power trends are investigated in recent years, and they are formulated as

$$y_t = \sum_{j=0}^{p} \beta_j L_t^{\theta_j} + u_t, \tag{1.18}$$

for $t = 1, 2, \ldots, T$, where in the work of Phillips (2007), L_t is a smoothly and slowly varying (SSV) function of time and $\theta_j = j$, while in the work of Robinson (2012b), $L_t = t$ and θ_j are unknown real values with $\theta_j > -1/2$.

Gao *et al.* (2020) considered a model with both a parametric global trend and a nonparametric local trend formulated as

$$y_t = g(\tau_t) t^{\theta_0} + e_t, \tag{1.19}$$

where $g(\cdot)$ is an unknown nonparametric function and $\theta_0 \geq 0$ is an unknown parameter. As in the nonparametric literature, the component $g(\cdot)$ is assumed to be bounded and smoothly varying, it captures a nonlinear trend of a quite varied nature. At the same time, the global trend part t^{θ_0} allows the outcome variable to increase without bounds as the horizon lengthens. The authors test the null hypothesis of $\mathbb{H}_0 : \theta_0 = 0$ against $\mathbb{H}_1 : \theta_0 > 0$. They also considered the hypothesis that $\mathbb{H}_0 : g(\tau)$ is a constant against $\mathbb{H}_1 : g(\tau)$ is not a constant. If one fails to reject either of these null hypotheses, everything goes back to some well-studied models, such as the nonparametric time trend model $y_t = g(\tau_t) + e_t$, when $\theta = 0$, and the power law model $y_t = ct^{\theta_0} + e_t$, when $g(\tau_t) = c$ is a constant.

1.4 The Mysteries of Trend: An Example

When practitioners know the actual form of the trend in the nonstationary time series, the analysis and interpretation of the underlying trending behavior seem straightforward. However, in empirical analysis, the real trending form is unknown, and uncovering precise trending forms from observed sequences of time series is challenging. The effectiveness of the unit-root tests introduced earlier is somewhat limited as they usually yield conflicting and inconclusive results. In essence, discerning between a unit root process showcasing stochastic trending behaviors and a trend-stationary process containing nonlinear deterministic trends, particularly in cases involving finite samples, proves to be a difficult task.

In this section, we provide a simulated example to illustrate this phenomenon. Specifically, the time series in the four subfigures plotted in Figure 1.4 are generated by the following steps:

(1) In Figure 1.4(a), x_{1t} is a simulated pure random walk process without drift $x_{1t} = x_{1t-1} + v_t$, where $v_t \overset{i.i.d.}{\sim} \mathcal{N}(0, 1)$ are standard normal distributions for $t = 1, 2, \ldots, 400$. As the process contains a stochastic trend, one may get different paths of the time series sequence for each realization. Without loss of generality, we focus on one realization of the random walk process as plotted in Figure 1.4(a).

(2) Now suppose that the true data generating process of x_{1t} is unknown, and we approximate x_{1t} using a polynomial function of time. The order of time polynomials needs to be properly chosen so as to capture trending behaviors in the time series and to ensure that the residuals are stationary. Alternatively, one can also use nonparametric methods to fit a time trend. As a simple illustration, we estimate the unknown coefficients in the polynomial function $x_{1t} = \sum_{j=0}^{6} \alpha_j t^j + e_t$ for $t = 1, 2, \ldots, 400$ and obtain the least squares estimates of $\widehat{\alpha}_j$ for $j = 1, 2, \ldots, 6$. The spuriously estimated time trend is then approximated by $\widehat{g}(t) = \sum_{j=0}^{6} \widehat{\alpha}_j t^j$, which is plotted as the dot-dashed curve in Figure 1.4(b).

(3) We generate a stationary AR(1) process by $u_t = 0.7u_{t-1} + \eta_t$ for $u_0 = 0$ and $\eta_t \overset{i.i.d.}{\sim} \mathcal{N}(0, 4)$, for $t = 1, 2, \ldots, 400$. The solid line in Figure 1.4(c) shows a combined process of two parts by $x_{2t} = \widehat{g}(t) + u_t$, where $\widehat{g}(t)$ is the fitted polynomial time trend from the realized unit-root process in the previous step. By construction, x_{2t} is a nonlinear trend-stationary process.

(4) In Figure 1.4(d), both the unit root process x_{1t} and the nonlinear trend-stationary process x_{2t} are plotted together for comparison. We also plot their first-order differences in Figures 1.4(e) and (f).

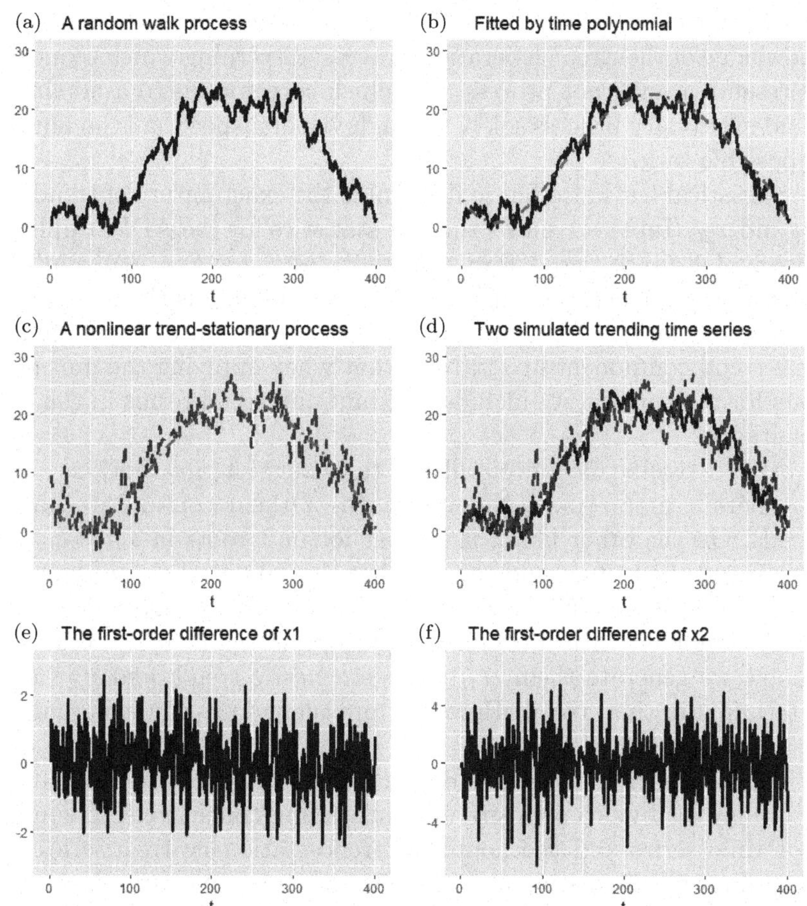

Figure 1.4. Simulated examples of time series with sample size $T = 400$: (a) A simulated random walk process $x_{1t} = x_{1,t-1} + v_t$ (black solid line), where $v_t \overset{i.i.d.}{\sim} \mathcal{N}(0,1)$, for $t = 2, \ldots, T$; (b) the fitted time polynomial $\widehat{g}_t$ (dot-dashed line) by least squares method; (c) a simulated nonlinear trend-stationary process $x_{2t} = \widehat{g}_t + u_t$ (dashed line), where $\widehat{g}_t$ is the estimated time polynomial in x_t, and $u_t = 0.7u_{t-1} + \eta_t$ in which $\eta_t \overset{i.i.d.}{\sim} \mathcal{N}(0,4)$ for $t = 2, \ldots, T$; (d) the two simulated trending time series x_{1t} and x_{2t} for comparison; (e) the first-order difference of x_{1t}; (f) the first-order difference of x_{2t}.

In Figure 1.4(d), the two trending time series exhibit striking similarities in their evolving directions and trending paths, despite their different underlying natures — one characterized by stochastic trends and the other by deterministic trends. In a scenario where a practitioner is presented with these time series without any prior knowledge of their real generation process, discerning which sequence represents a unit root process and which corresponds to a nonlinear trend-stationary process solely through visual inspection is an almost impossible task.

Statistical tests also fail to determine the exact nature of trends in x_{1t} and x_{2t}. Table 1.1 shows the test statistics for the ADF, Phillips–Perron, DF-GLS, and KPSS unit-root tests, for x_{1t}, x_{2t}, and the detrended sequences $x_{1t} - \widehat{g}_t$ and $x_{2t} - \widehat{g}_t$, where $\widehat{g}_t$ is the fitted time polynomial for the random walk process x_{1t} and also the nonlinear time trend component in x_{2t}. Note that when applying the unit root tests for x_{1t} and x_{2t}, we include constant and trend terms in the test equations.

At 5% significance level, all the tests suggest that both x_{1t} and x_{2t} contain unit roots.[1] Namely, both of them contain stochastic trends. On the other hand, if we test for unit roots in $x_{1t} - \widehat{g}_t$ and $x_{2t} - \widehat{g}_t$, all the tests would suggest that their detrended counterparts are stationary. In other words, both x_{1t} and x_{2t} are stationary time series combined with a nonlinear time trend $\widehat{g}_t$. In fact, according to how we generate them, x_{1t} has a stochastic trend, and x_{2t} has a deterministic time trend. However, prevalent unit-root tests fail to distinguish between them.

To conclude, the test results in Table 1.1 show that unit-root tests may lead us to opposite and misleading conclusions. The unit-root time series and the nonlinear trend-stationary time series are sometimes hardly distinguishable. Therefore, complex trends are full

Table 1.1. Unit-root test results.

	Test statistic				Critical values		
	x_{1t}	x_{2t}	$x_{1t} - \widehat{g}_t$	$x_{2t} - \widehat{g}_t$	10%	5%	1%
ADF test	–0.572	–2.527	–4.970	–8.124	–3.133	–3.421	–3.981
PP test	–0.572	–1.683	–5.077	–8.139	–3.313	–3.421	–3.981
DF-GLS test	–0.457	–2.620	–3.371	–6.420	–2.570	–2.890	–3.479
KPSS test	0.546	0.545	0.056	0.087	0.119	0.146	0.216

[1]Note that the results depend on how the stochastic trend in x_{1t} is realized in one simulation. The results in Table 1.1 are based on one realization of the simulated trending time series.

of mysteries (Phillips, 2010), and we may never understand their real nature by relying on unit-root tests.

1.5 Detrending Methods and Their Problems

As mentioned, many well-known and widely used time series models require the underlying data to be stationary. Otherwise, we may suffer from the problem of spurious regression. However, the presence of trends in the time series, whether deterministic or stochastic, violates this stationary requirement. This makes direct applications of classical time series regression models challenging. Economists often solve this problem by utilizing "detrending" techniques. They aim to transform nonstationary time series data into a stationary one, thereby meeting the crucial conditions of stationarity.

Some of the most commonly used detrending approaches are summarized and discussed in the work of Canova (1998). The author finds that various detrending methods lead to different patterns of estimated trends and cycles. In this section, we first review some of the popular detrending methods and then discuss their advantages and disadvantages when applying them in practice:

(1) *Fitting and removing linear or nonlinear functions of time*
When a trending pattern is seen in the time series plot, fitting a linear or nonlinear trend to the data is one of the simplest methods of detrending. For instance, we assume that only deterministic trends are present in the trending time series, then we approximate the data sequence by a function of time and then remove the estimated time trend to obtain a stationary process. In particular, suppose a pth-order ($p \geq 1$) polynomial trend exists in x_t:

$$x_t = \sum_{k=0}^{p} \alpha_k t^k + e_t, \tag{1.20}$$

where the polynomial order p is set in advance and e_t is a stationary process with zero mean and finite variance. We regress x_t over $(1, t, t^2, \ldots, t^p)$ and obtain the residuals $\widehat{e}_t = x_t - \sum_{k=0}^{p} \widehat{\alpha}_k t^k$ as the detrended version of x_t, where $\widehat{\alpha}_k$ are the OLS estimates of α_k for $k = 0, 1, \ldots, p$. In most of the scenarios, we only include a simple linear time trend and let $p = 1$. In general, however, we can always find $p \geq 1$ such that $\widehat{e}_t$ is close to a stationary process. If the polynomial order p in equation (1.20) is not determined beforehand, information criteria or shrinkage methods can be used to select an optimal value of p. Furthermore, even if the deterministic time trend is not a polynomial function of time, we can still employ this method, provided that we can show that the real-time trend function in x_t can be well approximated by polynomials of time. Of course, other nonlinear functions, such as those with breaks, are also useful if the existence of one or more structural breaks over time is justified in the time series.

(2) *Taking difference*
Taking difference is another commonly used method to eliminate trends in the nonstationary time series.[2] In essence, the differencing operation is a linear filter that allows high-frequency information to pass and filters out low-frequency information, such as long-term trends and cycles with relatively long periods.

For example, for an integrated process of order d (denoted by $I(d)$) with d being a positive integer, taking the difference of x_t for d times removes stochastic trends and gives a stationary process. That is, if $x_t \sim I(d)$, then $\Delta^d x_t \sim I(0)$, where Δ is the differencing operator and $I(0)$ represents a stationary process. For time polynomial trend of order p, differencing the time series for p times may also eliminate the trend component, though we may introduce moving average unit roots in the detrended series. However, when the trending mechanism is unknown, it may be dubious to remove trends by differencing the data.

(3) *The Hodrick and Prescott (HP) filter*
The HP filter method was proposed in Hodrick and Prescott (1997), and it is widely applied by macroeconomists. The HP filter separates the trend and cycle components by solving the optimization problem:

$$\min_{g_1,\ldots,g_t,\ldots,g_T} \sum_{t=1}^{T} (x_t - g_t)^2 + \lambda \sum_{t=2}^{T-1} \left((g_{t+1} - g_t) - (g_t - g_{t-1})\right)^2, \tag{1.21}$$

[2]Here, we only consider time series data. Sometimes, differencing for the cross-sectional data is quite useful in practice; see the first chapter by Yatchew (2003).

where $\{g_t\}_{t=1}^T$ is the trend sequence to be estimated. The smoothness of the estimated trend depends on the tuning parameter λ. In practice, λ takes different values for different frequencies of the time series. For example, $\lambda = 100, 1600$, and 14400 for the yearly, quarterly, and monthly data, respectively. As shown in Hodrick and Prescott (1997) and Hamilton (2018), the solution to equation (1.21) is an explicit linear function of all the observations of x_t for $t = 1, 2, \ldots, T$.

The HP filter method is widely applied in practice, however, it has several drawbacks. De Jong and Sakarya (2016) and Sakarya and De Jong (2020) discovered that there could still be nonstationary behaviors at the start or end of the filtered data. Hamilton (2018) found that the HP filter produces spurious autocorrelations in the estimated cycle series. Such dynamics are purely an artifact of the filter and have nothing to do with the real generating process of the data. Phillips and Jin (2021) showed that the HP filter may not successfully remove the trend component even if the time series is only $I(1)$ for commonly encountered sample sizes in practice.

(4) *Beveridge and Nelson's procedure*
Suppose that L is a lag operator and a nonstationary trending time series follows an ARIMA$(p, 1, q)$ process:

$$\Phi(L)(1 - L)x_t = \mu + \Theta(L)e_t, \tag{1.22}$$

where $\Phi(L) = \sum_{j=1}^p \phi_j L^j$, $\Theta(L) = \sum_{j=1}^q \theta_j L^j$, and e_t is an *i.i.d.* sequence with zero mean and variance $\sigma^2 > 0$. Beveridge and Nelson (1981) proposed a decomposition method that first transforms the expression to

$$(1 - L)x_t = \mu^* + C(L)e_t, \tag{1.23}$$

where $\mu^* = \Phi^{-1}(L)\mu$ and $C(L) = \Phi^{-1}(L)\Theta(L) = \sum_{j=0}^{\infty} c_j L^j$ with the coefficients $\sum_{j=0}^{\infty} j|c_j| < \infty$. Then, the nonstationary time series could be decomposed as

$$x_t = x_0 + \mu^* t + C(1)\sum_{s=1}^{t} e_s + C^*(L)e_t, \tag{1.24}$$

where $x_0 + \mu^* t$ is the deterministic trend, $C(1)\sum_{s=1}^{t} e_s$ is the stochastic trend, and $C^*(L)e_t$ is the stationary component in which $C^*(L) = \sum_{j=0}^{\infty} c_j^* L^j$ and $c_j^* = -\sum_{s=1}^{\infty} c_{j+s}$. The decomposition could be achieved based on consistent estimations of the parameters μ, ϕ_j, and θ_j in the ARIMA$(p, 1, q)$ model or c_j in the ARIMA$(0, 1, \infty)$ model. Removing the estimated deterministic and stochastic trends yields a detrended stationary sequence.

(5) *A regression method*
To fix the problems with the *Hodrick and Prescott* filter, Hamilton (2018) proposed a simple but useful regression method to separate trends from the nonstationary time series data. The method regresses $y_{t+h} (h \geq 1)$ on a constant and four most recent values of y_t by

$$y_{t+h} = \beta_0 + \beta_1 y_t + \beta_2 y_{t-1} + \beta_3 y_{t-2} + \beta_4 y_{t-3} + v_{t+h}, \quad (1.25)$$

where the fitted values of $\widehat{y}_{t+h} = \widehat{\beta}_0 + \widehat{\beta}_1 y_t + \widehat{\beta}_2 y_{t-1} + \widehat{\beta}_3 y_{t-2} + \widehat{\beta}_4 y_{t-3}$ are taken as trends and the estimated residuals $\widehat{v}_{t+h} = y_{t+h} - \widehat{y}_{t+h}$ are constructed as the transient (cycle) component. This simple method could be employed for a broad class of underlying processes, such as those with random walks or deterministic time trends.

Some other useful detrending methods are also discussed in the literature, for example, Canova (1998). For example, the frequency domain method and those using the unobserved component model, the one-dimensional index model, the model of common deterministic trends, and the model of common stochastic trends. Different detrending methods may result in different separations between trends and cycles.

Although these detrending methods are widely applied in empirical analysis, we may still need to be cautious about the problems and defects of these methods in practice. A commonly seen problem is that taking the first difference may not remove all the trend components as one may expect. We present an example related to the logarithm of quarterly aggregate disposable personal income,[3] which we denote as y_t.

[3]The data are downloaded from https://fred.stlouisfed.org/series/DPI.

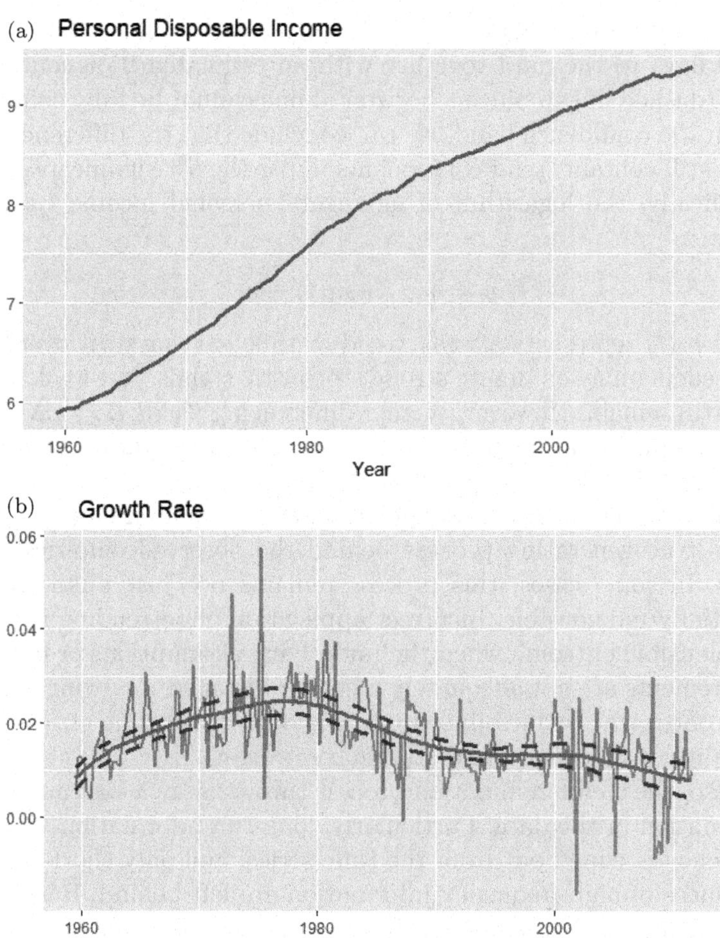

Figure 1.5. The trending time series of the logarithm of aggregate personal income in panel (a) and its first difference (i.e., growth rate) in panel (b). For the difference sequence, we fit a time trend (solid line) and estimate the 95% confidence band (dashed lines) using nonparametric kernel estimation methods.

Figure (1.5) shows an example to illustrate the problem of taking the difference. Panel (a) of the figure plots the quarterly records of disposable personal income, which is apparently a trending time series. Panel (b) shows the first difference of the trending time series as an attempt to remove embedded time trends. This obtains the growth rates as plotted in Panel (b). Using the kernel estimation method, a nonparametric regression against time fits a time trend (solid line) to the data together with an estimated 95% confidence band (dashed lines). Since a horizontal line cannot be fully contained within the confidence band, we can conclude that the difference time series still contain trend components under the 5% significance level. Specifically, the logarithm of aggregate personal income (y_t) may follow

$$y_t = y_{t-1} + m(t) + v_t, \tag{1.26}$$

where $m(t)$ represents a weak trend or time-varying drift over time. Differencing may eliminate strong stochastic trends, but weak trends may still remain. However, further differencing yields $\Delta_t^y = \Delta m(t) + \Delta v_t$ introducing a moving average unit root in the data and a weak trend $\Delta m(t)$ still exists in $\Delta^2 y_t$.

Another significant challenge arises from the often elusive nature of the true generating process behind the observed empirical time series. In some cases, this process remains not just unknown but essentially unknowable. Incorrect application of detrending methods is a plausible outcome when the underlying assumptions or essential requirements are not adequately met. For instance, a moving average unit root might be introduced into the differenced time series when, in reality, it follows a trend-stationary process.

Also, the trend-elimination process throws away a vast amount of information in the data. Particularly, long-run information with low frequency is wiped out from the time series, and only short-run disturbances of high-frequency information are left behind. Rao (2010) argued that the regressions using the differenced variables are useless to verify economic theories as they only reflect the relationship between short-run variables rather than the long-term equilibrium relationships. As shown by an example in the work of Cochrane (2012), the relationship between the original nonstationary data is quite significant, while the scatter plot turns out to show little correlation between the stationary sequences after differencing. Moreover, the time series sometimes needs to be differenced more than once before achieving stationarity when the trend component is quite complex. Eventually, little information is maintained in the stationary time series after differencing twice or even more times.

Furthermore, as the true data-generating process is unknown most of the time, misuse of the detrending methods usually leads to severe statistical problems. Nelson and Kang (1981) suggested that if we remove an estimated polynomial trend from an $I(1)$ process, we may introduce pseudo-periodic behavior in the detrended series. Therefore, the regression results based on the modified stationary sequences make no sense as we have artificially introduced the autocorrelations. Moreover, for a nonlinear trend-stationary process, one may need to take the difference more than once to obtain a stationary process. For example, in a scenario where $x_t = t^2 + v_t$ and v_t is a stationary $I(0)$ process, the second-order difference of x_t contains moving average unit roots, leading to an uninvertible process ($\Delta^2 x_t = 2 + \Delta^2 v_t$). For the time series with both global and local trends as $x_t = a + bt + g(\tau_t) + v_t$, where $g(\tau_t)$ is a local trend and v_t is stationary, differencing only eliminates the strong global trend $a + bt$, but the local trend $\Delta g(\tau_t)$ still remains in Δx_t even though its magnitude is notably smaller than that of the global trend $g(\tau_t)$.

To conclude, the methods used to eliminate trends can inadvertently remove a significant amount of valuable information in the data, an issue likened to "throwing the baby out with the bathwater" as described in the work of Cochrane (2012). Consequently, substantial information losses within trending time series can result in inefficiencies and inconsistencies in estimating crucial parameters of interest. The transformation also alters the focus of the regression analysis from a long-run relationship to a short-run correlation. Moreover, these improperly applied methods can give rise to various statistical complications, leading to intricate econometric challenges. To address these drawbacks, econometricians are proposing new regression models that directly tackle nonstationary trending time series instead of analyzing their stationarized alternatives. This serves as a primary motivation underlying the content of this book.

1.6 Summary

Nonstationary time series often exhibit trends that dominate their evolution over time. These trends can be either deterministic or stochastic, and the exact nature is usually unknown to researchers. Unit-root tests are commonly used tools to assess stationarity in time series data. However, these tests may struggle to distinguish between complex trends, especially when only a finite span of data is available and nonlinear time trends are permitted in the test regressions. Detrending methods like differencing or filtering are often used to remove trends in time series data. However, in the case of complex trends, these methods may be inappropriate and problematic as they can eliminate important information from the nonstationary time series.

Despite the development of econometric models for investigating trends, there is still a lack of sufficient and powerful tools to study complex trends in time series data. To address these challenges, researchers may need to explore alternative methods and models that can effectively capture and analyze complex trends in nonstationary time series. This could involve the development of more sophisticated econometric models and advanced statistical techniques tailored to handle the intricacies of nonstationary data with complex trends.

1.7 Bibliographical Notes

For the detailed methods and applications of nonlinear and nonstationary time series analysis, one can also refer to the works of Priestley (1988), Hamilton (1994), Hargreaves (1994), Clements and Hendry (1999), Fan and Yao (2003), Gao (2007), Tsay and Chen (2018), and Mills (2019). In particular, the works of Pfaff (2008) and Woodward *et al.* (2017) provide application examples with R codes.

To estimate and forecast trends in nonstationary time series, some other useful methods can also be applied in practice in addition to nonparametric kernel methods. For example, one can use the orthogonal series methods as in the works of Gallant (1981) and DeVore and Lorentz (1993), the ARIMA models as in the work of Box *et al.* (1987), the filtering methods as in the work of Hodrick and Prescott (1997) and Gomez (2001), the wavelet methods as in the works of Andreas and Treviño (1997) and Lineesh and John (2010), the exponential smoothing method as in the work of Hyndman *et al.* (2008), the smoothing splines as in the works of Hyndman *et al.* (2005) and Morton *et al.* (2009), and neural networks as in the work of Zhang and Qi (2005). Note that practitioners have to be cautious about certain prerequisites when applying such methods.

Chapter 2

Trending Time Series Models

> When the idea of cointegration was developed, over a decade later, it became clear immediately that if a pair of series was cointegrated then at least one of them must cause the other.
>
> — Granger (2003)

In practice, it is crucial to establish the long-term association between economic variables, such as testing specific economic or scientific theories. For example, validating the permanent income hypothesis (PIH) may necessitate examining the long-run relationship between aggregate income and consumption. In these instances, direct regression of the nonstationary time series becomes essential, given that regressing their differenced versions primarily reflects short-term correlations only.

Meanwhile, detrending converts nonstationary time series to stationary alternatives poses significant challenges, particularly when the underlying trend's nature is not well understood. Detrending potentially leads to the loss of crucial information, and inappropriately detrending data can trigger econometric issues, including wrong regression relations and biased coefficient estimates, consequently distorting the results of statistical analyses and forecasts.

These aspects underscore the necessity of developing trending regression models, which accommodate situations where the regressor or the regressand is nonstationary and trending, and where the regressor and error term may also exhibit endogenous correlation. By accounting for these complexities, trending regression models effectively capture the trending patterns and reveal the long-term dynamics between economic variables to offer valuable insights for economic and scientific analyses.

As discussed in Section 1.3, trends in nonstationary time series could be either stochastic or deterministic. The literature regarding trending time series models could be divided into two major streams. One line focuses on modeling deterministic trends, and the other involves stochastic trends. Note that among the econometric practice of testing for common trends, cointegration analysis is the most prevalent. Researchers are interested in finding common stochastic trends when the nonstationary time series are regarded as integrated time series based on unit root tests.

This chapter is organized as follows. We start by introducing regression models with deterministic trends, which can be of linear and nonlinear, parametric and nonparametric forms. We then proceed to cointegration — a key concept in modern time series econometrics including linear, nonlinear, and fractional cointegrations. Before introducing the application examples of trending time series regressions in economics and finance in Sections 2.7 and 2.8, we discuss the problem of spurious regression and the problem of endogeneity that are frequently encountered in practice.

2.1 Regression Models with Deterministic Trends

A simple linear time trend model can be established as

$$y_t = a + bt + e_t, \tag{2.1}$$

for $t = 1, 2, \ldots, T$, where e_t is usually assumed to be a stationary process with zero mean and finite variance. The unknown coefficients a and b could be easily estimated, for example, by the ordinary least squares (OLS) method that $\widehat{b} = \sum_{t=1}^{T}(t-\bar{t})(y_t-\bar{y})/\sum_{t=1}^{T}(t-\bar{t})^2$ and $\widehat{a} = \bar{y} - \widehat{b}\bar{t}$, where $\bar{y} = \sum_{t=1}^{T} y_t/T$ and $\bar{t} = (T+1)/2$. As t grows with sample size T, $\widehat{b}$ converges to b at the rate of $O_p(T^{-3/2})$, much faster than $O_p(1/\sqrt{T})$. In many of the empirical analyses, the conventional t-test is applied to examine the presence of a linear time trend, i.e., researchers would like to see the significance of the slope coefficient b. For example, when y_t denotes the annual global mean temperatures from 1901 to 2020, then the value of b reflects the average warming rate per year over the past 120 years, and a significantly positive value of b indicates compelling statistical evidence supporting global warming. In terms of appropriate estimation and testing of b, however, one has to take into account possible serial correlation and conditional heteroskedasticity in the error term e_t. The serial correlation (persistence) of e_t may severely affect the performance of the usual t-test especially when the sample size is not sufficiently large.

Canjels and Watson (1997) studied the estimation and inference problems in the linear trend model of equation (2.1). In particular, the error term follows an autoregressive process $e_t = \rho e_{t-1} + u_t$, where ρ, the initial value u_1, and the variance of u_t are unobserved. They find that the asymptotic distribution of the estimators depends on the local-to-unity parameter ρ that measures the persistence in the errors and the variance of the initial error term.

Bunzel and Vogelsang (2005) proposed tests for hypotheses on the parameters of the deterministic trend function of a univariate time series. Their tests do not require prior information on serial correlation in the data, and they are robust to strong serial correlation. Even when the data contain a unit root, the tests still have the correct size asymptotically. Also see the works of Harvey *et al.* (2007, 2009); Perron and Yabu (2009) for different testing approaches. Recently, Elliott (2020) revisited the testing problem of b when there is a potentially strong serial correlation in e_t. The paper developed new tests for the hypothesis $\mathbb{H}_0 : b = 0$ against $\mathbb{H}_1 : b > 0$, and the author concentrates on the size and power properties of the tests for models where the serial correlation has a unit root ($\rho = 1$) or near unit root ($\rho \approx 1$).

However, the functional form of deterministic trends may not necessarily be linear. As pointed out in the work of Harvey (1997), "... unless the time period is fairly short, these trends cannot be adequately captured by straight lines. In other words, a deterministic linear time trend is too restrictive ..." Therefore, a straightforward extension of equation (2.1) is the one with a kth-order time polynomial formulated as

$$y_t = \sum_{j=0}^{k} c_j t^j + e_t, \tag{2.2}$$

for $j \geq 1$ and $t = 1, 2, \ldots, T$, where e_t is covariance stationary. When there exist conditional heteroskedasticity and serial correlation in the error term, Grenander and Rosenblatt (1954) showed that the OLS estimators of c_j's are asymptotically equivalent to the generalized least squares (GLS) estimators, and they also have the same variance (i.e., efficiency) when T tends to infinity. Another form of nonlinear time trend is a piecewise linear trend

$$y_t = a + bt + c(t - \gamma)1_{(t>\gamma)} + e_t, \tag{2.3}$$

where the breakpoint location $t = \gamma$ is known or unknown and needs to be estimated (Perron, 1989). Indeed, multiple breaks are also allowed to occur in such kind of piecewise linear deterministic trends.

In recent years, there has been an increasing number of regression models that incorporate deterministic trends with nonlinear and nonparametric forms. In this context, we discuss several significant examples of time series and panel data models that include nonparametric time trends as part of the model. The most basic nonparametric model for a nonlinear trend-stationary time series is formulated as

$$y_t = g(\tau_t) + e_t, \tag{2.4}$$

for $t = 1, 2, \ldots, T$, where $\tau_t = t/T$, $g(\cdot)$ is an unknown smooth function defined at equally spaced fixed-design points on $[0, 1]$, and the error term e_t is usually assumed to be stationary. The nonparametric function fits the time series of y_t with smoothness and flexibility that captures the trends or time-varying mean (i.e., low-frequency information) in y_t. As no prior functional forms are imposed on $g(\cdot)$, we are free from the risk of model misspecification. Nevertheless, the drawback is the lack of empirical interpretation as no parameters are included in (2.4) for reference. As demonstrated by Robinson (1989, 1991), the nonparametric function $g(\cdot)$ now depends on the sample size T, and this kind of "standardization" is necessary to provide asymptotic justification for nonparametric smoothing estimators. The intuitive behind this "intensity" assumption is that when sample size tends to infinity, it is an increasingly intense sampling of data points on $[0, 1]$. By accumulating information, it enables us to obtain a consistent estimation of the unknown smooth function of $g(\cdot)$.

The model could be estimated by, for example, the nonparametric local constant kernel estimation method (Fan and Gijbels, 1996; Li and Racine, 2007, etc.). Specifically, at the local area of τ_t, we approximate the nonparametric trend function $g(\tau_t)$ by a constant value c and establish the objective function $\sum_{s=1}^{T}(y_s - c)^2 K_{ts}$, where $K_{ts} = K\left(\frac{s-t}{Th}\right)$, $K(u)$ is the kernel function, which is usually a probability density function defined on $[0, 1]$, and h is the bandwidth parameter that controls the smoothness of the estimated time trend. The solution minimizing the objective function gives the nonparametric local constant kernel estimator for $g(\tau_t)$ as

$$\widehat{g}(\tau_t) = \frac{\sum_{s=1}^{T} K_{ts} y_s}{\sum_{s=1}^{T} K_{ts}}. \tag{2.5}$$

After obtaining $\widehat{g}(\tau_t)$ for $t = 1, 2, \ldots, T$, the residuals $\widehat{e}_t = y_t - \widehat{g}(\tau_t)$ could be regarded as the detrended values of y_t. The local constant estimator in (2.5) is also called the Nadaraya–Watson (NW) estimator. Robinson (1997) provides the asymptotic theories for this model, where the error term e_t is assumed to be serially dependent on short-range, long-range, or negative type. The local linear kernel estimation method is similar to the local constant method but approximates $g(\tau_t)$ by $c_0 + c_1(\tau_s - \tau_t)$ in the local area of τ_t. Therefore, the local linear estimator is $\widehat{g}(\tau_t) = \widehat{c}_0$, where $(\widehat{c}_0, \widehat{c}_1)$ minimizes the objective function $\sum_{s=1}^{T}(y_t - c_0 - c_1(\tau_s - \tau_t))^2 K_{ts}$.

Juhl and Xiao (2005) considered a semiparametric trending time series model of the form

$$y_t = f(\tau_t)'\theta(\tau_t) + u_t, \tag{2.6}$$

for $t = 1, 2, \ldots, T$, where $\tau_t = t/T$. In the model, $f(\tau_t)$ is a vector of prespecified deterministic functions; for example, when $f(\tau_t) = 1$, the model becomes $y_t = \theta(\tau_t) + \varepsilon_t$, nesting our simplest nonparametric model of equation (2.4). Then, testing for constancy of $\theta(\tau_t)$ is equivalent to testing whether the mean of y_t is a constant value or time-varying. If $f(\tau_t) = (1, \tau_t)'$, the model presents a linear trend form while the intercept or the slope is allowed to change smoothly over time. Meanwhile, $\theta(\tau_t)$ is a vector of unknown, possibly time-varying parameters. The stochastic component u_t follows an AR(p) process $u_t = \sum_{i=1}^{p} \rho_i u_{t-i} + \varepsilon_t$, where the roots of $1 - \rho_1 z - \rho_2 z^2 - \cdots - \rho_p z^p = 0$ lie outside the unit circle, ensuring that u_t is a stationary process.

The null hypothesis stands for the trivial case when there are no changing trends, i.e., $\mathbb{H}_0 : \theta(\tau_t) = \theta_0$, that the coefficients are constants. Juhl and Xiao (2005) emphasized that since the test is nonparametric, they have power against very general alternatives.

To test for the null hypothesis, Juhl and Xiao (2005) borrowed the ideas in the nonparametric conditional moment tests by Zheng (1996); Fan and Li (1999); Lavergne and Vuong (2000); Hsiao and Li (2001) and constructed a test statistic $J_T = Th^{1/2}\widehat{I}_T/\widehat{\Sigma}^{1/2}$, where $\widehat{I}_T = (1/T^2h)\sum_{t=1}^{T}\sum_{s\neq t}^{T} K_{ts}\widehat{\varepsilon}_t\widehat{\varepsilon}_s$. In the expression of $\widehat{I}_T$, $K_{ts} = K\left(\frac{\tau_t-\tau_s}{h}\right)$, and $\widehat{\varepsilon}_t$ is estimated from the AR(p) process of u_t that $\widehat{\varepsilon}_t = \widehat{u}_t - \sum_{i=1}^{p}\widehat{\rho}_i\widehat{u}_{t-i}$, and $\widehat{u}_t = y_t - f(\tau_t)'\widehat{\theta}$, and $\widehat{\theta}$ is the obtained based on the OLS regression of y_t on $f(\tau_t)$ under the null hypothesis. To guarantee consistent estimation of $\widehat{\rho}_i$'s under both the null and the alternative hypotheses, they are estimated by regressing $\tilde{y}_t$ on $(\tilde{y}_{t-1}, \tilde{y}_{t-2}, \ldots, \tilde{y}_{t-p})$, where $\tilde{y}_t$ is the detrended values of y_t that $\tilde{y}_t = y_t - \widehat{g}(\tau_t)$. Furthermore, $\widehat{\Sigma} = (2/T^2h)\sum_{t=1}^{T}\sum_{s\neq t}^{T} K_{ts}^2\widehat{\varepsilon}_t^2\widehat{\varepsilon}_s^2$ is a consistent estimator of the variance of $\widehat{I}_T$, and scaling by which the test statistic J_T follows a standard normal distribution under the null hypothesis.

Gao and Hawthorne (2006) investigated a semiparametric partially linear model formulated as

$$y_t = x_t'\beta + g(\tau_t) + e_t, \tag{2.7}$$

for $t = 1, 2, \ldots, T$, where y_t is a trending time series, x_t is a k-dimensional vector of stationary explanatory variables, $g(\cdot)$ is an unknown smooth function defined on [0,1] that captures the trending behaviors in y_t, and e_t is the error term that represents stationary disturbances. According to the usual profile estimation method for partially linear models (Härdle *et al.*, 2000), the regression coefficient vector β in equation (2.7) could be estimated by the least squares method with respect to a transformed equation of $y_t^* = x_t^*\beta + e_t$, where y_t^* and x_t^* are the detrended values of y_t and x_t based on model (2.4). With $\widehat{\beta}$, a consistent estimator of β, the unknown function $g(\tau_t)$ could be estimated afterward as the time trend in $y_t - x_t'\widehat{\beta}$ using nonparametric estimation methods. Since no functional forms are imposed on $g(\tau_t)$, it is estimated in a data-driven manner that captures the trending information in y_t without the risk of model misspecification.

Following a similar way to the testing scheme as in the work of Juhl and Xiao (2005), Gao and Hawthorne (2006) proposed an adaptive test for certain parametric specifications of $g(\tau_t)$, for example, $\mathbb{H}_0 : g(\tau_t) = \alpha_0 + \gamma_0 t$ against $\mathbb{H}_1 : g(\tau_t) \neq \alpha + \gamma t$ for some $\theta_0 = (\alpha_0, \gamma_0) \in \Theta$ and all $\theta = (\alpha, \gamma) \in \Theta$, where Θ is the parameter space in $\mathbb{R}^2$. Specifically, the test statistic is $L^* = \max L_T(h)$ over $h \in H_T$, the set of bandwidths, where $L_T(h) = \sum_{t=1}^{T} \sum_{s=1,\neq t}^{T} K_{ts}\tilde{e}_s\tilde{e}_t / \tilde{S}_T$, and $\tilde{S}_T^2 = 2\sum_{t=1}^{T} \sum_{s=1}^{T} K_{ts}^2 \tilde{e}_s^2 \tilde{e}_t^2$ is the variance of the numerator in $L_T(h)$, $\tilde{e}_t$ is the least squares residual from the regression model under $\mathbb{H}_0$ that minimizes the objective function $\sum_{t=1}^{T} (y_t - x_t'\beta - \alpha - \gamma t)^2$. In general, the specification of the trend form in $\mathbb{H}_0$ and $\mathbb{H}_1$ can be any parametric function of t indexed by the parameter vector θ, such as the polynomial time trends.

Cai (2007) proposed a time-varying coefficient trending time series model

$$y_t = \alpha(\tau_t) + x_t'\beta(\tau_t) + e_t, \tag{2.8}$$

for $t = 1, 2, \ldots, T$, where $e_t = \sigma_t(x_t)u_t$ is allowed to be heteroscedastic with $\mathrm{E}(e_t^2|x_t) = \sigma_t^2(x_t)$, and $\{x_t, u_t\}$ is assumed to be strictly stationary α-mixing. In this model, the regression coefficients have much flexibility to change smoothly over time. They are estimated in a data-driven manner without the risk of model misspecification. This time-varying coefficient model nests some special cases, for example, the model reduces to the usual constant coefficient model when $\alpha(\tau_t) \equiv \alpha$ and $\beta(\tau_t) \equiv \beta$, and to model (2.7) when $\beta(\tau_t) \equiv \beta$. But in terms of estimation, this model is easier to estimate than the semiparametric model (2.7). That is, one only has to approximate the unknown coefficient functions $\alpha(\tau_t)$ and $\beta(\tau_t)$ by linear functions of τ_t at each local point of $\tau \in [0, 1]$.

Gao and Robinson (2016) investigated a semiparametric model that incorporates deterministic trends as a moving mean of y_t after fractional filtering. The model is formulated as

$$\xi_t(L; \delta, \theta)\ y_t = g(\tau_t) + e_t, \tag{2.9}$$

where $\xi_t(L; \delta, \theta)$ is a parametric filter incorporating fractionally differencing with short memory corrections with unknown fractional and short memory parameters δ and θ. The nonstationary time series y_t contain stochastic trends, which are removed by the fractional differencing operator $\xi_t(L;\delta,\theta)$. However, the deterministic trend could not be fully eliminated after fractional differencing. The nonparametric function $g(\tau_t)$ may exhibit a mild pattern of trend or time-varying mean, but it forms a complex trend in y_t via integration — the inverse process of fractional differencing.

Dong and Linton (2018) studied the additive nonparametric models with time variable and both stationary and nonstationary regressors as

$$y_t = \beta(\tau_t) + g(z_t) + m(x_t) + e_t, \tag{2.10}$$

where $\beta(\cdot), g(\cdot), m(\cdot)$ are smooth functions with unknown forms, z_t is a stationary process, while x_t is an integrated process, $\beta(\tau_t)$ captures the deterministic trend component in y_t. This model provides a flexible approach to explain the dependent variable y_t using deterministic and possibly nonlinear time trends, and stationary and nonstationary variables. All unknown functions in equation (2.10) are estimated by the series method, which gives an explicit solution for the estimators obtained by the OLS. In particular, the approximations are designed as follows. $\beta(r) = \sum_{j=0}^{\infty} c_{1,j}\phi_j(r)$, where $\{\phi_j(r)\}$ is an orthonormal basis in the Hilbert space $L^2[0,1]$; $g(z) = \sum_{j=0}^{\infty} c_{2,j}p_j(z)$, where $\{p_j(z)\}$ is an orthonormal basis in $L^2(V, dF(z)) = \{q(z) : \int_V q^2(z)dF(z) < \infty\}$ in which $F(z)$ is a distribution on the support V that may not be compact; $m(x) = \sum_{j=0}^{\infty} c_{3,j}\mathscr{H}_j(x)$, where $\{\mathscr{H}_j(x)\}$ are Hermite polynomials that are orthogonal with respect to the density $\exp(-x^2)$ in the Hilbert space $L^2(\mathbb{R})$. By setting the truncation parameters k_1, k_2, and k_3 for the three expansions respectively, equation (2.10) becomes a simple linear regression model with unknown parameters $\{c_{1,j}\}_{j=0}^{k_1}$, $\{c_{2,j}\}_{j=0}^{k_2}$, $\{c_{3,j}\}_{j=0}^{k_3}$ to be estimated by least squares method. The three nonparametric functions can then be estimated by $\widehat{\beta}(r) = \sum_{j=0}^{k_1} \widehat{c}_{1,j}\phi_j(r)$, $\widehat{g}(z) = \sum_{j=0}^{k_2} \widehat{c}_{2,j}p_j(z)$, and $\widehat{m}(x) = \sum_{j=0}^{k_3} \widehat{c}_{3,j}\mathscr{H}_j(x)$.

Dong *et al.* (2021) argued that additivity in equation (2.10) is a strong assumption ruling out interaction effects and can be violated by some datasets. Therefore, they proposed a nonparametric regression model

$$y_t = m(\tau_t, z_t, x_t) + e_t, \tag{2.11}$$

where z_t is a stationary process, x_t is an integrated process, and the nonparametric function $m(\cdot)$ is defined on $[0,1] \times V_z \times \mathbb{R}$ in which V_z is the support of z_t. They employ the tensor product $\{\phi_i(r)\} \otimes \{p_j(z)\} \otimes \{h_l(x)\}$ as the orthonormal basis that $\mathcal{B}_{ijl}(r,z,x) = \phi_i(r)p_j(z)h_l(x)$, where $\phi_i(r)$, $p_j(z)$, and $h_l(x)$ are basis functions defined as those in Dong and Linton (2018). Then, the regression function in (2.11) can be well approximated by $m(r,z,x) = \sum_{i,j,l=0}^{\infty} c_{ijl}\mathcal{B}_{ijl}(r,z,x)$, where the coefficients are the inner products $c_{ijl} = \iiint_{[0,1]\times V\times\mathbb{R}} m(r,z,x)\mathcal{B}_{ijl}(r,z,x)F(z,x)drdzdx$ in which $F(z,x)$ is a density function defined on $V \times \mathbb{R}$. After properly defining the truncation parameters k_1, k_2, and k_3, the unknown coefficients c_{ijl}, for $i = 0,1,\ldots,k_1, j = 0,1,\ldots,k_2$, and $l = 0,1,\ldots,k_3$, can be directly estimated by the least squares method. Finally, the regression function $m(\tau,z,x)$ can be recovered by $\widehat{m}(r,z,x) = \sum_{i=0}^{k_1}\sum_{j=0}^{k_2}\sum_{l=0}^{k_3} \widehat{c}_{ijl}\mathcal{B}_{ijl}(r,z,x)$.

Gao *et al.* (2020) included both a parametric global trend and a nonparametric local trend in a model of the form

$$y_t = g(\tau_t)t^{\theta_0} + e_t, \tag{2.12}$$

where $g(\cdot)$ is an unknown nonparametric function associated with a power function of t in which $\theta_0 \geq 0$ is an unknown parameter. The nonparametric component $g(\cdot)$ is bounded and smoothly varying over time, and it captures nonlinear trends of a quite varied nature. At the same time, t^{θ_0} is a global trend that allows the outcome variable to increase without bounds as the horizon lengthens. However, the authors find that either the global or the local profile method fails in estimating θ_0 consistently as it disappears from the leading term of the objective function for profile methods. As a consequence, a new objective function $R_T(\theta) = \left[\lambda_T \cdot \ln\left(T^{-1}\sum_{t=\lfloor Th\rfloor+1}^{T} \tau_t^{2\theta}\widehat{g}(\tau_t,\theta)\right)^2\right]^2$ has to be proposed to obtain consistent estimation for θ_0, where $\lambda_T = 1/\ln T$ and $\widehat{g}(u,\theta) = \sum_{t=1}^{T} t^{\theta}y_tK_h(u-\tau_t)/\sum_{t=1}^{T} t^{2\theta}K_h(u-\tau_t)$ is the profile least squares estimator in which $K_h(u) = K(u/h)/h$ and $K(\cdot)$ is the usual kernel function.

In panel data models, nonlinear and nonparametric time trends are also incorporated, estimated, and tested in certain models to capture trending behaviors. For example, Robinson (2012b) investigated a simple separated structure for panel data by considering a nonparametric trending regression with cross-sectional dependence formulated as

$$y_{it} = \alpha_i + \beta(\tau_t) + e_{it}, \tag{2.13}$$

for $i = 1, 2, \ldots, N, t = 1, 2, \ldots, T$, where α_i denotes individual-specific effect, while $\beta(\tau_t)$ denotes a common trend over time, the error terms e_{it} are unobservable zero-mean random variables, uncorrelated and homoscedastic across time but possibly correlated and heteroscedastic over the cross-section. Unlike the time-fixed effects in the usual panel data models, $\beta(\tau_t)$ is a smooth function of τ_t. By imposing the restriction $\sum_{i=1}^{N} \alpha_i = 0$ or $\sum_{i=1}^{N} \omega_i \alpha_i = 0$, $\sum_{i=1}^{N} \omega_i = 1$, for some vector of weights $\omega = (\omega_1, \ldots, \omega_N)'$, one can transform the panel data model to a time series model as equation (2.4) by removing all the α_i's through taking the simple average or weighted average of equation (2.13). The paper provides a thorough discussion of the asymptotic properties of the nonparametric kernel estimator for $\beta(\tau)$.

Li *et al.* (2011) established a nonparametric time-varying coefficient panel data model of the form

$$y_{it} = f(\tau_t) + x_{it}'\beta(\tau_t) + \alpha_i + e_{it}, \tag{2.14}$$

for $i = 1, 2, \ldots, N, t = 1, 2, \ldots, T$, where $x_{it} = (x_{it,1}, \ldots, x_{it,k})'$ is a k-dimensional vector of stationary regressors and α_i denotes individual fixed effects when it is correlated with x_{it} through an unknown correlation structure. Therefore, the model allows for the situation that the dependent variable y_{it} contains a common trend $f(\tau_t)$ for all $i = 1, 2, \ldots, n$, and the marginal effects of x_{it} on y_{it} also vary smoothly over time. To estimate the time-varying coefficients $f(\tau)$ and $\beta(\tau)$, the paper proposes two methods. The first method takes averages of equation (2.14) over i using the condition $\sum_{i=1}^{N} \alpha_i = 0$ and transforms the panel data model to a varying-coefficient time series regression model $y_{\cdot t} = f(\tau_t) + x_{\cdot t}'\beta(\tau_t) + e_{\cdot t}$. Then, both $f(\tau)$ and $\beta(\tau)$ can be estimated by applying the nonparametric local linear kernel estimation methods by Fan and Gijbels (1996) and Cai (2007). Due to information loss in the averaging process, the rate of convergence of the averaged local linear estimate of $\beta(\tau)$ is only $O_p(1/\sqrt{Th})$. That is, large N does not contribute to the efficient estimation of $\beta(\tau)$. To get an estimate that has a faster rate of convergence, the paper further proposes a local linear dummy variable approach. In particular, the individual fixed effects are retained in the regression model by using dummy variables, and the time-varying coefficients are estimated via a profile least squares approach as in the works of

Su and Ullah (2006) and Sun *et al.* (2009). As a consequence, the convergence rate of $\widehat{\beta}(\tau)$ becomes $O_p(1/\sqrt{NTh})$ as more observed individuals in the data provide additional information that enhances the efficiency of the nonparametric estimator for the time-varying coefficients.

To allow the regressors to be nonstationary trending rather than stationary, Chen *et al.* (2012) proposed a semiparametric trending panel data model with cross-sectional dependence

$$y_{it} = x_{it}'\beta + f(\tau_t) + \alpha_i + e_{it}, \tag{2.15}$$

$$x_{it} = g_t + x_i + v_{it}, \tag{2.16}$$

for $i = 1, 2, \ldots, N, t = 1, 2, \ldots, T$, where both $f(\tau_t)$ and $g(\tau_t)$ are nonparametric time trend functions defined on [0, 1]. Note that x_{it} now exhibits a similar data-generating structure as that in equation (2.13), while the regression model is similar to that by Li *et al.* (2011) except for β being a vector of constant coefficients over time. The unknown coefficients β, $f(\tau)$, and α_i are estimated via a profile least squares method after employing dummy variables to express fixed effect terms in the model.

Zhang *et al.* (2012) focused on the problem of whether y_{it} share a common trend and investigated the trending panel data model:

$$y_{it} = \alpha_i + x_{it}'\beta + f_i(\tau_t) + e_{it}, \tag{2.17}$$

for $i = 1, 2, \ldots, N, t = 1, 2, \ldots, T$, where $f_i(\tau_t)$ are nonlinear and nonparametric time trends for y_{it}, respectively. They developed an R^2-based test to examine the null hypothesis $\mathbb{H}_0 : f_i(\tau) = f(\tau)$ for $\tau \in [0, 1]$ assuming that y_{it}, for all i, share the same trend. In the first step of the test, one has to obtain the estimated residuals under the null hypothesis. That is, estimate equation (2.15) by Chen *et al.* (2012) and compute the residuals by $\widehat{e}_{it} = y_{it} - x_{it}'\widehat{\beta} - \widehat{f}(\tau_t)$. Then, it suffices to test for time trends in $\widehat{e}_{it}$, for $i = 1, 2, \ldots, N$ to see whether the estimated common trend $\widehat{f}(\tau_t)$ could represent all the trending components in y_{it}.

Robinson (2012a) studied a spatial lattice regression model for observations over time, space, or space-time. Let $u = (u_1, u_2, \ldots, u_d)'$ be a d-dimensional vector of multi-indices for $d \geq 1$ and p_i denote the number of power law terms for index u_i. The model is formulated as

$$y_u = \sum_{i=1}^{d} \sum_{j=1}^{p_i} \beta_{ij} u_i^{\theta_{ij}} + e_u, \quad u \in \mathbb{Z}_+^d, \tag{2.18}$$

for $u_i = 1, 2, \ldots, n_i$, $i = 1, 2, \ldots, p$, where β_{ij} and θ^{ij} are the unknown parameters, the error term e_u is covariance stationary with zero mean, and $\mathbb{Z}_+ = \{j : j = 0, 1, \ldots\}$. To estimate the unknown parameters β_{ij} and θ^{ij}, the paper employs the nonlinear least squares estimation (NLSE) method by $\arg\min_{\beta_{ij}, \theta_{ij}} \sum_u (y_u - \sum_{i=1}^{d} \sum_{j=1}^{p_i} \beta_{ij} u_i^{\theta_{ij}})^2$ and derives the asymptotic properties for $\widehat{\theta}_{ij}$ and $\widehat{\beta}_{ij}$.

In a special case, $d = 2$ and index $u_1 = 1, 2, \ldots, N$ denotes spatial points while index $u_2 = 1, 2, \ldots, T$ denotes time, model (2.18) becomes a spatial-temporal model. The model further reduces to a simple "power law" model for time series when $d = 1$. That is,

$$y_t = \sum_{j=1}^{p} \beta_j t^{\theta_j} + e_t, \tag{2.19}$$

for $t = 1, 2, \ldots, T$, where both the powers θ_j's and the corresponding coefficients β_j's are unknown. For estimation and testing purposes, we require $\theta_j > -1/2$ for all j, and the error term e_t must be zero mean and covariance stationary with short memory. For $\theta_j < -1/2$, β_j would not be estimable as the trending signal is too weak and therefore disguised by noise. For $\theta_j = -1/2$, β_j is estimable, however, not suitable for inference as the central limit theorem requires θ_j to lie in the interior of a compact set.

2.2 Cointegration: Common Stochastic Trends

Besides deterministic time trends, many empirical data are believed to contain stochastic trends, such as the unit root time series. For example, the aggregate income and consumption, the logarithms of stock prices, the gross national product, and so on. In practice, economic theory often suggests that a certain subset of variables shall be governed by a stable long-run equilibrium relationship, and certain economic forces would restore equilibrium once the evolution of these variables drifts away from the long-run relationship. Therefore, empirical economists need econometric models and methods to examine the hypothesis from these economic theories based on economic time series data.

Cointegration models offer a useful framework for modeling and testing these long-run equilibrium relationships between unit root time series. These models reflect common stochastic trends. In the case of cointegrated time series, linear combinations of these nonstationary variables become stationary, indicating that any deviations from this long-run equilibrium relationship are transitory and will eventually be corrected. Cointegration models have widespread applications in econometrics, finance, and other fields that deal with unit root time series data. Note that in this context, econometric analysis are conducted directly on the nonstationary data instead of their differenced counterparts, which are stationary but only reflect the short-run correspondences between the time series.

The concept of *cointegration* was first introduced by Granger (1981, 1983) and Granger and Weiss (1983) and then systematically studied by Engle and Granger (1987). The representation, estimation, and testing of cointegration models are investigated in these papers in which a formal definition of cointegration was given as follows:

> the components of the vector x_t are said to be cointegrated of order d, b, denoted $x_t \sim CI(d, b)$, if all the components of x_t are $I(d)$, and there exists a vector $\alpha (\neq 0)$ so that $z_t = \alpha' x_t \sim I(d-b), b > 0$. The vector α is called the cointegration vector.

Specifically, a combination of $I(1)$ processes x_t becomes a stationary $I(0)$ process z_t, which rarely drifts far from zero if it has zero mean and will often cross the zero line. Therefore, equilibrium will occasionally occur. Engle and Granger (1987) also concentrated on the special case when $d = b = 1$ and provided a testing procedure for the cointegrating relationship based on the work of Dickey and Fuller (1979).

A well-known example of cointegration in economics is the theory of Purchasing Power Parity (PPP). The idea behind PPP is that in the absence of transportation costs and other frictions, identical goods and services should have the same price in two countries. If this is not the case, there is an opportunity for arbitrage. For example, let P_t denote the index of the price level in the US (in US dollars), P_t^* denote the price index in the UK (in pounds), and S_t the exchange rate between the currencies (USD per GBP). Then, the

PPP theory simply suggests that $P_t = S_t P_t^*$. Taking logarithm, we have the equation $p_t = s_t + p_t^*$, where $p_t = \ln(P_t)$, $s_t = \ln(S_t)$, and $p_t^* = \ln(P_t^*)$. In reality, however, PPP is a theoretical concept, and various factors such as transportation costs, trade barriers, and differences in the quality of goods can hinder immediate equalization of prices at every time t. However, in a broader sense, the PPP relation is expected to roughly hold. As a result, a less strict version of the PPP hypothesis is that the variable z_t defined by $z_t = p_t - s_t - p_t^*$ is stationary with zero mean, even though the original time series p_t, s_t, and p_t^* are all $I(1)$. In this example, the cointegrating vector is $(1, -1, -1)'$, based on which the trends in p_t, s_t, and p_t^* cancel out and thus create a stable long-run relationship.

We can express and then estimate the cointegration relationship using a regression model. Specifically, we select one variable as the dependent variable, and the others as regressors:

$$y_t = \alpha + x_t'\beta + e_t, \tag{2.20}$$

for $t = 1, 2, \ldots, T$, where x_t and y_t are integrated time series of order 1 and e_t is a stationary $I(0)$ process. In other words, the stochastic trends in x_t and y_t cancel out in the linear combination of $y_t - x_t'\beta$. The establishment of the cointegration relationship in equation (2.20) indicates the long-run relationship between x_t and y_t, and β is also known as the long-run response of x_t to y_t. But how do they respond in the short run so as to maintain the equilibrium relationship in the long run?

To answer this question, an error-correction model (ECM) incorporates short-term dynamics could be established as

$$\Delta y_t = \gamma_0 + \sum_{j=1}^{p} \gamma_{1j} \Delta y_{t-j} + \sum_{j=0}^{q} \gamma_{2j} \Delta x_{t-j} + \delta \widehat{e}_{t-1} + v_t, \tag{2.21}$$

for $t = 2, 3, \ldots, T$, where $\widehat{e}_t = y_t - \widehat{\alpha} - x_t'\widehat{\beta}$ are the regression residuals of equation (2.20). The coefficient δ measures the adjustment speed on how y_t responds to the equilibrium error in the previous period. The implication of ECM is that when two or more integrated time series are cointegrated, there is some adjustment mechanism that corrects back to equilibrium when deviations from the long-term equilibrium occur. This mechanism prevents the disequilibrium errors in the long-run relationship from growing larger and larger by accumulation.

Since y_t is artificially selected as the dependent variable, ECM can also be established for other variables in the cointegration system. For example, the ith element in x_t

$$\Delta x_{it} = \gamma_{0i} + \sum_{j=1}^{p} \gamma_{1j,i} \Delta y_{t-j} + \sum_{j=0}^{q} \gamma_{2j,i} \Delta x_{i,t-j} + \delta_i \widehat{e}_{t-1} + v_{it}, \quad (2.22)$$

for each i, respectively. Similarly, δ_i denotes whether x_{it} responds to the equilibrium error at time $t-1$ as well as the speed of adjustment. The significance of δ and δ_i's allows for Granger causality testing, which helps determine the direction of causality between variables.

To justify the existence of cointegration, many statistical tests have been developed in the literature. The commonly used tests include the Engle and Granger (1987) test and the Phillips and Ouliaris (1990) test based on a single-equation model. It is a two-step residual-based test by Engle and Granger (1987) that one has to first estimate equation (2.20) and then employ the Dickey–Fuller (DF) test to test for unit roots in the residuals. If reject the null hypothesis of the DF test, we can conclude that there is no unit root in the residuals, and x_t and y_t are cointegrated. Note that due to the reason that the cointegrating coefficients are estimated with uncertainty in the first step, the critical values of the tests differ from the usual DF tests, and they depend on the number of regressors in x_t. Johansen (1991) tested for more than one cointegrating relationship between multiple unit root time series. Breaks are also considered in cointegration tests, for example, the Gregory and Hansen (1996) test for cointegration with a single structural break, the Hatemi-J (2008) test for cointegration with two structural breaks, and the Maki (2012) test for cointegration with multiple structural breaks.

The cointegration relationship between multivariate integrated time series could be expressed in a VAR(p) model as

$$Y_t = \Phi D_t + \Pi_1 Y_{t-1} + \cdots + \Pi_p Y_{t-p} + u_t, \quad (2.23)$$

for $t = 1, 2, \ldots, T$, where $Y_t = (y_{1t}, \ldots, y_{kt})'$, and D_t contains deterministic terms. If the equation $|I_k - \Pi_1 z - \cdots - \Pi_p z^p| = 0$ has a unit root, then some or all of the variables in Y_t are $I(1)$ and may be cointegrated. We then write the VAR(p) model as

$$\Delta Y_t = \Phi D_t + \Pi Y_{t-1} + \Gamma_1 \Delta Y_{t-1} + \cdots + \Gamma_p \Delta Y_{t-p} + u_t, \tag{2.24}$$

where $\Pi = \Pi_1 + \cdots + \Pi_p - I_k$ and $\Gamma_j = -\sum_{s=j+1}^{p} \Pi_s$, for $j = 1, 2, \ldots, p-1$. Let $r = rank(\Pi)$. If $r = 0$, then Y_t could not form a cointegration relationship. Otherwise, if $0 < r < k$, then Y_t is $I(1)$ with r linearly independent cointegrating vectors.

In the following few years after the introduction of this important concept, cointegration quickly became popular and widely applied in various empirical analyses. The estimation and testing theory of linear cointegration was also extensively studied and developed. For example, Stock (1987) studied the asymptotic properties of the OLS and NLS estimators of the cointegrating vectors. Stock and Watson (1988) developed a test for the number of common stochastic trends. Phillips and Ouliaris (1988, 1990) employed the principal components and the residual-based methods to test for cointegration, respectively. Likelihood-based estimation and inference methods for cointegration regressions are established and applied by Phillips (1991), Johansen (1988, 1991, 1995), and Johansen and Juselius (1990, 1992) with error correction representations.

2.3 Nonlinear Cointegration

The classical cointegration models characterize linear associations between integrated time series. In a broader perspective, the cointegration relationships can also be nonlinear. Therefore, nonlinear cointegration provides a more general framework that captures nonlinear associations between integrated time series. Specifically, for a vector of integrated time series (x_t, y_t) where y_t is a scalar $I(1)$ process and x_t is a $k \times 1$ vector of $I(1)$ processes, there has been a class of nonlinear cointegration models as follows.

The study of nonlinear cointegration initiates from the seminal papers by Park and Phillips (1999, 2001). When the nonlinear structure of a long-run equilibrium relationship between scalar integrated time series x_t and y_t is specified based on certain economic theory, a parametric nonlinear cointegration model is formulated as

$$y_t = f(x_t; \theta) + u_t, \tag{2.25}$$

for $t = 1, 2, \ldots, T$, where $f : \mathbb{R} \times \mathbb{R}^m \to \mathbb{R}$ is a known function in which θ is an m-dimensional vector of unknown parameters and u_t is assumed to be a martingale difference sequence. The usual estimation method of θ is the nonlinear least squares (NLS) approach that $\widehat{\theta} = \arg\min_{\theta \in \Theta} Q_T(\theta)$, where Θ is the parameter set and $Q_T(\theta)$ is the loss function that $Q_T(\theta) = \sum_{t=1}^{T}(y_t - f(x_t; \theta))^2$.

Chang *et al.* (2001) incorporated nonlinear time trends and allowed for both stationary and integrated regressors. They consider the nonlinear regression model given by

$$y_t = g(t, \pi_0) + f(w_t, \alpha_0) + m(x_t, \beta_0) + u_t, \tag{2.26}$$

for $t = 1, 2, \ldots, T$, where w_t and x_t are stationary and integrated regressors, respectively, functional forms of g, f, and m are known but with unknown parameters π_0, α_0, and β_0 to be estimated. They find that the convergence rates of the NLS estimators in functions of integrated regressors depend on the nature of the nonlinear function. When $m(\cdot)$ is an integrable function, the convergence rate of the NLS estimator of β_0 is $\sqrt[4]{T}$, slower than the usual rate of $\sqrt{T}$ for the NLS estimates in the standard stationary regressions. On the other hand, the convergence rates of the NLS estimator of β_0 are faster than $\sqrt{T}$ when function $m(\cdot)$ is an asymptotically homogeneous function with increasing asymptotic orders.

Bae and De Jong (2007) studied the US long-run money demand function using the nonlinear cointegration model:

$$y_t = \beta_0 + \beta_1 \ln|x_t| + u_t, \tag{2.27}$$

where both x_t and y_t are integrated time series and u_t is stationary but serially correlated rather than a martingale difference sequence as in the works of Park and Phillips (2001) and Chang *et al.* (2001). Also, u_t and Δx_t are likely to be correlated. To allow for the presence of serial correlation and possible endogeneity in the error term, they proposed a fully modified nonlinear cointegration least square (NCLS) estimator. Their theory shows that the estimator of β_1 is $\sqrt{T}$-consistent rather than T-consistent.

Chang and Park (2011) considered the nonlinear regression with integrated regressors that are contemporaneously correlated with the regression error. Whether or not the limit distribution is affected by the presence of endogeneity, however, depends upon the functional type of the parameter derivative of the regression function. If it is asymptotically homogeneous, the limit distribution of the nonlinear least squares estimator has an additional bias term reflecting the presence of endogeneity. On the other hand, the endogeneity does not have any effect on the nonlinear least squares limit theory if the parameter derivative of regression function is integrable.

Chan and Wang (2015) also studied the univariate nonlinear parametric cointegration under a general framework that established the weak consistency of the NLS estimator. The framework is also applicable for various nonstationary time series, including partial sums of linear processes and Harris recurrent Markov chains. Furthermore, they allowed the error term to be serially correlated, and cross-dependent on the regressor that causes the problem of endogeneity. They show that the limit distribution of the NLS estimator under the endogeneity situation is different from that with martingale error structures.

In recent decades, specification tests for nonlinear cointegration have also been developed that enable researchers to test for certain parametric specifications. Testing for such a common trend relationship would be a statistical verification of the corresponding economic or financial theory. Gao *et al.* (2009) proposed a nonparametric kernel test for nonlinear nonstationarity against nonlinear stationarity. Choi and Saikkonen (2010) developed tests for the null hypothesis of cointegration in the nonlinear regression model with $I(1)$ variables. Hong and Phillips (2010) modified the RESET test to obtain power against both nonlinear cointegration and no cointegration alternatives to assess the adequacy of a linear cointegrating relation against certain forms of nonlinear cointegration and the alternative of no cointegration. Wang and Phillips (2012) provided a general theory of specification tests that is applicable for a wider class of nonstationary regressors that includes both unit root and near unit-root processes. Dong and Gao (2018) proposed tests for the specification of nonlinear cointegration models using the nonparametric series method with the presence of endogeneity where the regressor is allowed to be fractionally integrated. Wang *et al.* (2018) developed a test of parametric specification in a nonlinear cointegrating regression model by using the marked empirical processes without needing to select a bandwidth parameter.

In practice, however, the true cointegration coefficient (vector) may be time-varying, and a misspecified constant coefficient (vector) can hardly capture the changes in the cointegration relationship. This is particularly the case when the time series covers a considerably long period of time. Hansen (1992) and Quintos and Phillips (1993) established the Lagrange multiplier tests to examine the parameter consistency in the cointegration models, where the regression coefficient is assumed to follow a Gaussian random walk process when parameter constancy is violated.

To avoid the problem of model misspecification, nonparametric models are also established for nonlinear cointegration relationships. The model is formulated as

$$y_t = f(x_t) + u_t, \tag{2.28}$$

for $t = 1, 2, \ldots, T$, where function f is unknown and not restricted to certain parametric forms. Normally, we require that $f(\cdot)$ be a smooth function.

The nonparametric cointegration model was initially studied by Wang and Phillips (2009a,b). The model could be estimated by nonparametric kernel or sieve methods based on observed values of $\{x_t, y_t\}_{t=1}^T$. Wang and Phillips (2016) established nonlinear cointegrating regression models in which the regressor is allowed to be endogenous and driven by long memory innovations, and the regression errors are also allowed to be serially dependent. Li *et al.* (2016) studied parametric nonlinear regressions under the Harris recurrent Markov chain framework and found that the convergence rates for the estimators rely not only on the properties of the nonlinear regression function but also on the number of regenerations for the Harris recurrent Markov chain. Hu *et al.* (2021) developed an asymptotic theory for nonlinear cointegrating power function regression, which allows for both endogeneity and heteroskedasticity. For a detailed review on nonlinear cointegration, see the work of Tjøstheim (2020).

Since the regression coefficient may not be a constant over a long span of the time period, to allow for possible changes in the coefficients, Park and Hahn (1999) developed the cointegration regression model with time-varying coefficients of the form

$$y_t = x_t'\beta_t + e_t, \tag{2.29}$$

for $t = 1, 2, \ldots, T$, where x_t is a k-dimensional vector of $I(1)$ processes and $\beta_t = \beta(\tau_t)$ for $\tau_t = t/T$ is a smooth function defined on $[0, 1]$ representing the smoothly changing coefficients. Phillips *et al.* (2017) also considered the same model by nonparametric kernel methods. However, they find that the usual asymptotic methods of kernel estimation break down when the functional coefficients are multivariate. They explain this breakdown as a kernel-induced degeneracy in the weighted signal matrix associated with the nonstationary regressors, which is a new phenomenon in the kernel regression literature. See the works of Bierens and Martins (2010) and Li *et al.* (2020) for more detailed discussions.

We may also allow for functional coefficients in nonlinear cointegration models. Specifically, the cointegrating coefficient vector β is allowed to vary over a random variable z_t that

$$y_t = \beta(z_t)' x_t + u_t, \tag{2.30}$$

where $\beta(\cdot)$ is a k-dimensional vector of coefficients depending on z_t. Contributions to the literature include the works of Cai *et al.* (2009) and Xiao (2009) for proposing this kind of model, and then the works of Sun *et al.* (2013) and Phillips and Wang (2021) for further developments.

Threshold cointegration (Cai *et al.*, 2017) restricted nonlinear cointegration to threshold type only where

$$\boldsymbol{X}_t = A\boldsymbol{X}_{t-1} 1(\boldsymbol{X}_{t-1} \in D) + B\boldsymbol{X}_{t-1} 1(\boldsymbol{X} \in D^c) + \boldsymbol{e}_t, \tag{2.31}$$

in which $\boldsymbol{X}_t = (x_{1t}, x_{2t})'$ is bivariate and D is a subset of $\mathbb{R}^2$. The model allows for stationary-like or cointegration-like behaviors in different regions of D and D^c.

Dong *et al.* (2017) considered a general model specification test for nonlinear multivariate cointegrating regressions in which the regressors consist of a univariate integrated time series and a vector of stationary time series. Their model is formulated as

$$y_t = m(x_t, z_t) + u_t, \tag{2.32}$$

where x_t is a nonstationary integrated time series, z_t is stationary, and the regressors and the errors are generated from the same innovations so that the model accommodates endogeneity.

The cointegration regression model could be semiparametric, i.e., it is partially parametric with nonparametric components in an additive manner that

$$y_t = g(x_t; \theta) + f(v_t) + u_t, \tag{2.33}$$

where function g is known with unknown coefficients θ and function f is unknown with integrated process v_t. In the work of Gao and Phillips (2013), the parametric component is assumed to be linear that $g(x_t; \theta) = \theta' x_t$, and x_t is assumed to be dependent with v_t through unknown functional forms. In the work of Kim and Kim (2012), the authors proposed a two-step estimation method for a partial parametric model with multiple integrated time series based on the decomposition of the nonparametric part of the regression function into homogeneous and integrable components.

Dong and Linton (2018) considered an additive form of nonlinear cointegration model that incorporates deterministic time trend, stationary and integrated processes by

$$y_t = \beta(\tau_t) + g(z_t) + m(x_t) + u_t, \tag{2.34}$$

where $\beta(\tau)$ is an unknown deterministic function defined on [0, 1], z_t is a stationary process, x_t is an integrated process, and u_t is the error term.

A double-nonlinear cointegration model is investigated by Lin *et al.* (2020) that could nest many popular models discussed in previous examples. The model is specified as

$$G(y_t, \beta_0) = g(x_t) + u_t, \tag{2.35}$$

where the dependent variable y_t, after a strictly increasing transformation specified by the parametric family $\{G(y, \beta) : \beta \in \theta\}$, is related to the univariate unit-root regressor x_t via an unknown link function $g : \mathbb{R} \to \mathbb{R}$.

2.4 Fractional and Other Forms of Cointegration

Almost at the same time, researchers became aware of the fractionally integrated time series, which is formally introduced by

Granger and Joyeux (1980), Granger (1980), and Granger (1981). They established the fractionally integrated process as

$$(1-L)^d x_t = u_t, \tag{2.36}$$

for $t = 1, 2, \ldots, T$, where u_t is a WN process. The fractional differencing operator $(1-L)^d$ is defined through the infinite series expansion:

$$(1-L)^d = \sum_{j=1}^{\infty}(-1)^j \binom{d}{j} L^j, \tag{2.37}$$

where $\binom{d}{j} = d(d-1)\cdots(d-j-1)/j!$.

The statistical properties of x_t depend on the value of d. When $d \geq 0.5$, x_t is nonstationary as its variance goes to infinity with t. When $|d| < 0.5$, the stochastic process x_t is stationary, and its autocorrelation function $\rho(j) \sim j^{2d-1}$, which decay in a hyperbolical manner with j. When $0 < d < 0.5$, the process x_t is said to exhibit *long-memory* as $\sum_{j=-T}^{T}|\rho(j)|$ tends to infinity as $T \to \infty$. Otherwise, x_t is a *short-memory* process when $-0.5 < d < 0$.

In most cases, fractionally integrated data could arise as a result of aggregation. Robinson (1994) proposed tests for unit root and other forms of nonstationarity against a certain class of alternatives including fractionally and seasonally fractionally differenced processes. To consistently estimate the memory parameter d with confidence intervals, Shimotsu and Phillips (2005) investigated an exact form of the local Whittle (LW) estimator which does not rely on differencing or tapering. The estimator is shown to be consistent with Gaussian limit distributions.

Then, fractional cointegration became a natural extension of the classical definition of cointegration. Gil-Alana (2003) proposed a two-step testing procedure of fractional cointegration in macroeconomic time series based on the work of Robinson (1994). Hualde and Robinson (2010) considered a semiparametric multivariate fractionally cointegrated system where integration orders are possibly unknown and I(0) unobservable inputs are assumed to exhibit nonparametric spectral density. For a detailed discussion on possible empirical applications, see the work of Gil-Alana and Hualde (2009). Johansen and Nielsen (2012, 2018) considered model-based inference in a fractionally cointegrated (or cofractional) vector autoregressive model (CVAR), based on the Gaussian likelihood conditional on initial values. They show that the likelihood ratio test statistic for the usual CVAR model is asymptotically chi-squared-distributed. The work is then extended by Johansen and Nielsen (2019) in which the authors relax critical moment conditions and incorporate the possibility that the cointegrating vectors are non-stationary. Zhang *et al.* (2019) proposed a model-free eigenanalysis method for identifying cointegrated components of nonstationary time series that allows the integration orders (either integers or fractionals) of the observable series to be unknown and to possibly differ. Andersen and Varneskov (2021) proposed the Local speCtruM (LCM) approach for joint significance of the regressors in standard predictive regressions where some of the variables may be fractionally integrated with different orders. Their LCM procedure is based on fractional filtering and band spectrum regression using a suitably selected set of frequency ordinates.

Unbalanced cointegration refers to the situation where the integration orders of the observables are different, but their corresponding balanced versions (with equal integration orders after filtering) are cointegrated in the usual sense. Hualde (2006) introduced the unbalanced cointegration that $y_t \sim I(\delta)$, while $x_t \sim I(\delta+\theta_n)$, where θ_n measures the gap between their integration orders. There are three possibilities for θ_n: (1) $\theta_n \equiv 0$ for the usual balanced cointegration case; (2) $\theta_n \to 0$ with n for the weakly unbalanced cointegration case; (3) $\theta_n = \theta \neq 0$ for the strongly unbalanced case. The weakly unbalanced cointegration is a situation of "near-cointegration", where the only difference concerning the "usual" fractional cointegration is that the orders of the two observable series differ in an asymptotically negligible way. In the strongly unbalanced case, a bivariate unbalanced cointegration is formulated as

$$y_t = \alpha + \beta\Delta^{\theta} x_t + \Delta^{-\gamma} u_{1t}, \tag{2.38}$$

$$x_t = \Delta^{-(\delta+\theta)} u_{2t}, \tag{2.39}$$

where $\delta > \gamma$ and $(u_{1t}, u_{2t})'$ is a zero mean $I(0)$ process. Therefore, y_t and x_t are not cointegrated, but y_t and the filtered series $\Delta^{\theta} x_t$ are cointegrated. Hualde (2014) investigated the asymptotic properties of nonlinear least squares estimators of the long-run parameters in a bivariate unbalanced cointegration framework. Within this setting, the long-run linkage between the observables is driven by both the cointegrating parameter and the difference between the integration orders of the observables, which we consider to be unknown.

Seasonal cointegration models are also proposed in the literature, where we analyze the time series at different frequencies in the spectrum. As defined by Hylleberg *et al.* (1990) and Engle *et al.* (1993), a time series is seasonally integrated of order d at frequency θ (denoted as $x_t \sim I_\theta(d)$) for $\theta \in (-\pi, \pi]$, if d is the smallest integer that

$$S(L)^d x_t = \eta_t, \tag{2.40}$$

where $S(L) = 1 - e^{i\theta} L$ and η_t is a covariance stationary process.[1] In particular, for x_t being an $I_\theta(1)$ process, its variance goes to infinity as $t \to \infty$, and innovations have permanent effects on the seasonal pattern of x_t. For example, a sequence of quarterly data $x_t \sim I_\theta(1)$ for $\theta = 0, \pi$, and $\pm\pi/2$ if

$$(1 - L^4)x_t = \eta_t, \tag{2.41}$$

where η_t is an $I(0)$ process. Suppose that all the components in the vector of $X_t = (x_{1t}, \ldots, x_{kt})'$ are $I_\theta(1)$, then x_t is said to be seasonally cointegrated at frequency θ if there exists a nonzero vector α such that $\alpha' X_t \sim I_\theta(0)$. Note that in this scenario, only the unit root at frequency θ is eliminated by cointegration, and unit roots at other frequencies may exist in $\alpha' X_t$. A stronger condition is that x_t is seasonally integrated of order 1 at some frequencies, not necessarily at the same frequency for all components. Then, X_t is said to be fully cointegrated if $\alpha' X_t$ is a stationary $I(0)$ process. In other words, all the unit roots are eliminated by cointegration.

The error correction model for seasonal cointegration is investigated in the work of Johansen and Schaumburg (1999) that provides the asymptotic distribution of the likelihood ratio test for the cointegrating rank. In the work of Lee (1992), testing procedures for cointegration and seasonal cointegration were developed for nonstationary time series which have unit roots at zero and seasonal frequencies using maximum likelihood inference. Several null

[1]A complete version is $S(L)^d \Phi(L) x_t = \eta_t$, where $\Phi(L)$ may have unit roots at zero or other seasonal frequencies different from θ.

hypotheses can be tested separately for each case of interest without any prior knowledge about the existence of cointegration relations at other frequencies. An empirical example is illustrated by using Canadian data on unemployment and immigration rates. Gregoir (2010) extended the framework of the fully modified OLS estimator introduced by Phillips and Hansen (1990) to the case of seasonally cointegrated processes at a given frequency. The paper proposes a test whose null hypothesis is the existence of seasonal cointegration and derives estimates of the cointegration vectors that allow for asymptotic normal inference.

2.5 Spurious Regressions

While cointegration reflects a long-term relationship between integrated time series, spurious regression occurs when a significant correlation or regression relationship is erroneously found between independent integrated time series. In the early twentieth century, researchers were puzzled by empirical findings revealing absurd associations between time series variables. At that time, people were unaware that these seemingly evident but unreasonable correlations actually originated from spurious regressions. For example, Yule (1926) found a puzzling high correlation of 0.95 between the proportion of church of England marriages to all marriages and the mortality rate over the period 1866–1911. He calls it "a nonsense correlation" as the share of marriages in the church of England should not be correlated with the mortality rate. Hendry (2000) found an extremely high correlation (0.998) between inflation and cumulative rainfall in the UK. Apparently, however, rainfall should neither cause nor predict inflation. Skog (1988) reported that the correlation between the quarterly index of intravenous drug abuse in Stockholm and Wolfer's index of sunspot activity is 0.91 (1965–1970), but it is impossible that sunspot activities cause human drug addiction. Before uncovering the mechanism of spurious regression in recent decades, economists were confused about such high correlations between these seemingly independent variables.

In terms of econometric analysis, spurious regression was first studied by Granger and Newbold (1974) with several simulation examples. To uncover the real nature of spurious regression, Phillips (1986) provided the large-sample asymptotic properties for the least square estimators in spurious regressions. Specifically, the author considers the linear regression $y_t = \alpha + x_t\beta + e_t$, for $t = 1, 2, \ldots, T$, where $x_t = x_{t-1} + w_t$, $y_t = y_{t-1} + v_t$ are both integrated time series in which v_t and w_t follow *i.i.d.* distributions with zero mean and variances σ_v^2 and σ_w^2, respectively. Then, as $T \to \infty$, the OLS estimator of $\widehat{\beta}$ has a nondegenerate limiting distribution and does not converge in probability to a constant. The conventional t-ratios diverge to infinity with T, causing a high rejection rate of $\mathbb{H}_0 : \beta = 0$ in the usual t-tests as the t-ratio exceeds the critical values (such as 1.96 under 5% significance level) when T is large. At the same time, low values for the Durbin–Watson statistic and moderate to high values of the coefficient of determination R^2 are to be expected in spurious regressions. These results all lead to incorrect inferential results using the conventional methods of regression analysis in applied economics.

To further understand spurious regression, Phillips (1998) made use of the general representation theory of a stochastic process in terms of an orthonormal system of basis functions. Specifically, trends themselves can be validly modeled in a variety of ways. This explains why significant regression coefficients occur in manifestly incorrect regression specifications relating variables that are statistically independent. In light of the representation theory, the author emphasized that cointegrating regressions do not explain trends. Instead, they only relate trends in multiple time series. Also, the nature of trending mechanisms in economics is little understood and econometricians have little guidance from economic theory models about meaningful economic specification.

When extended to nonparametric settings, Phillips (2009) considered the regression model $y_t = g(x_t) + e_t$ and found that all the usual characteristics of linear spurious regression are manifest in the context of local-level regression, including divergent significance tests, moderate to high R^2, and Durbin–Watson ratios converging to zero.

Recently, Chen and Tu (2019) explored the spurious effects in linear regressions with moderately explosive processes, where $y_t = \rho_{yn}y_{t-1} + u_{yt}$, $x_t = \rho_{xn}x_{t-1} + u_{xt}$ in which $\rho_{yn} = 1 + c_y/n^\alpha$ and $\rho_{xn} = 1 + c_x/n^\alpha$ with c_y and c_x being positive constants and $\alpha \in (0, 1)$. The paper derives the asymptotic properties of the least squares estimator of $\widehat{\beta}$ in $y_t = \alpha + x_t\beta + e_t$ and finds that they depend on the values of c_x and c_y. To obtain robust inference for spurious regressions with mildly explosive processes, the authors consider a balanced regression approach by augmenting the spurious regression model with a lagged dependent variable and lagged regressors as

$$y_t = \beta x_t + \lambda x_{t-1} + \gamma y_{t-1} + e_t \tag{2.42}$$

and estimate the unknown coefficients (β, λ, γ) by least squares method. Then the t-statistic regarding $\widehat{\beta}$ follows a standard normal distribution and is robust to the local parameters c_x and c_y. Similar work was done by Lin and Tu (2020) that considers spurious regression where x_t and y_t are generated from a more general mechanism in which $\rho_{yn} = 1 + c_y/k_n$ and $\rho_{xn} = 1 + c_x/k_n$, and $k_n = o(n)$ is a sequence increasing to infinity with n. They are called processes moderately deviated from a unit root (PMDUR) that includes the case of $k_n = n^\alpha$ for $\alpha \in (0, 1)$ as in the work of Chen and Tu (2019).

By allowing the coefficients to depend on certain covariates, Tu and Wang (2022) investigated the spurious regression in the functional-coefficient regression model:

$$y_t = \beta(z_t)x_t + u_t, \tag{2.43}$$

where $\beta(\cdot)$ is an unknown smooth function of some stationary covariate z_t, and x_t is a unit root process. The authors showed the asymptotic properties of the nonparametric local constant kernel estimator of $\widehat{\beta}(z)$ when both x_t and y_t are generated from independent random walks (i.e., the regression relationship is spurious). They also find that when x_t and y_t are not cointegrated, the global significance tests proposed by Xiao (2009) and Sun *et al.* (2016) are likely to fail and produce misleading conclusions for practitioners. Therefore, to test for spurious regression in functional-coefficient model regression models, they propose a semiparametric balanced regression, by augmenting regressors of the original spurious regression with lagged dependent variable and independent variables:

$$y_t = \beta(z_t)x_t + \lambda(z_{t-1})x_{t-1} + \phi y_{t-1} + e_t, \tag{2.44}$$

where lagged values of z_t, x_t, and y_t are incorporated in the model. Based on standard nonparametric inferential asymptotics, this model is able to detect spurious regression and is found robust to the true relationship between the integrated processes.

2.6 The Problem of Endogeneity

Apart from the issues posed by the ambiguity in the nature of trends, economists also encounter the challenges caused by the problem of endogeneity. This issue widely exists across many research domains and emerges when the explanatory variable and error term in a regression model are correlated. The resulting correlations can introduce bias and inconsistency in the estimators of coefficients or fitted models, even with a large sample size. They also hinder the correct estimation and testing of the actual relationship between dependent and independent variables. This problem is especially prevalent in economic analysis, given the intricate interconnections between various economic factors in the data.

Several reasons can cause the problem of endogeneity. First is the simultaneous causality between the dependent variable and regressors, for example, income and consumption, demand (supply) quantity, and price. The second possible reason is the omission of critical variables in regression models. In other words, the error term contains omitted variables that not only significantly affect the dependent variable but also strongly correlate with the explanatory variable. Other reasons may also cause the problem of endogeneity, for example, measurement errors in explanatory variables and biased sample selection. Endogeneity may cause conventional estimators or fitted models to become biased and inconsistent if proper correction methods are not employed.

In the subsequent chapters, one of our goals is to investigate the estimation and inference methods for the linear regression models with both problems of nonstationarity and endogeneity. Specifically, we consider the regression relationship between time series x_t and y_t in a linear regression model formulated as

$$y_t = \alpha + x_t'\beta + e_t, \tag{2.45}$$

for $t = 1, 2, \ldots, T$, where α is the intercept term, β is a $k \times 1$ vector of unknown coefficients, and the error term e_t is stationary. The regressor x_t is a $k \times 1$ vector of trending time series:

$$x_t = g_t + v_t, \tag{2.46}$$

in which g_t is a $k \times 1$ vector of trending components in x_t and the error vector v_t is also stationary. Note that the regressand y_t

is supposed to be trending time series as well unless $\beta \equiv 0$ or the combination of regressors $x_t'\beta$ does not contain trends.

The error terms e_t and v_t represent the shocks to x_t and y_t, respectively. They may be affected by certain common factors simultaneously. As a result, the stationary error terms e_t and v_t are highly likely to be correlated with an unknown structure. Consequently, in the model (2.45), the regression error e_t becomes correlated with at least one of the regressors in x_t, causing the problem of endogeneity in the regression model.

We now provide some theoretical justifications to address the consequences of the problem of endogeneity in a simple univariate[2] linear regression model $y_t = \alpha + \beta x_t + e_t$ for $t = 1, 2, \ldots, T$. The OLS estimator of β is defined as

$$\widehat{\beta} = \frac{\sum_{t=1}^{T}(x_t - \bar{x})(y_t - \bar{y})}{\sum_{t=1}^{T}(x_t - \bar{x})^2} = \beta + \frac{\sum_{t=1}^{T}(x_t - \bar{x})(e_t - \bar{e})}{\sum_{t=1}^{T}(x_t - \bar{x})^2}, \tag{2.47}$$

where $\bar{x} = T^{-1}\sum_{t=1}^{T} x_t$, $\bar{y} = T^{-1}\sum_{t=1}^{T} y_t$, and $\bar{e} = T^{-1}\sum_{t=1}^{T} e_t$. We consider two different cases:

(1) *The regressor x_t is purely stationary*
In this case, $g_t \equiv 0$ and $x_t = v_t$ in equation (2.46), and we assume that $\widehat{Q} = T^{-1}\sum_{t=1}^{T}(x_t - \bar{x})^2$ and $S_{xe} = T^{-1}\sum_{t=1}^{T}(x_t - \bar{x})(e_t - \bar{e})$. When T goes to infinity, we have $\widehat{Q} \to \sigma_v^2$ and $S_{xe} \to \sigma_{ev}$ almost surely. Note that σ_{ev} is a nonzero value when e_t and v_t are correlated due to the problem of endogeneity. Then, for the OLS estimator $\widehat{\beta} = \beta + S_{xe}/\widehat{Q}$, we have

$$\widehat{\beta} \to \beta + \frac{\sigma_{ev}}{\sigma_v^2}, \tag{2.48}$$

almost surely when sample size T tends to infinity, suggesting that $\widehat{\beta}$ is an inconsistent estimator for β as the bias term limit σ_{ev}/σ_v^2 does not dissipate with T.

(2) *The regressor x_t has a weak trend*
In this case, x_t is a nonstationary trending time series as formulated in equation (3.2). The trend g_t, however, is relatively weak that we could assume $T^{-1}\sum_{t=1}^{T}(g_t - \bar{g}_T)^2 \to S_g$,

[2]For simplicity, we only consider the univariate regression model. Similar results could be derived for multivariate regression models.

where $\bar{g}_T = T^{-1}\sum_{t=1}^{T} g_t \to \bar{g}$ and $S_g = \int_0^1 (g(\tau) - \bar{g})^2 d\tau$ for $0 < S_g < \infty$, and $\bar{g} = \int_0^1 g(\tau)d\tau$. Therefore, as T goes to infinity, $\widehat{Q} = T^{-1}\sum_{t=1}^{T}(x_t - \bar{x})^2$ converges to $S_g + \sigma_v^2$, and $S_{xe} = T^{-1}\sum_{t=1}^{T}(x_t - \bar{x})(e_t - \bar{e})$ converges to σ_{ev}. Then, for the OLS estimator $\widehat{\beta} = \beta + S_{xe}/\widehat{Q}$, we have

$$\widehat{\beta} \to \beta + \frac{\sigma_{ev}}{S_g + \sigma_v^2}, \tag{2.49}$$

almost surely with T. Since the bias term has a nonzero limit $\sigma_{ev}/(S_g + \sigma_v^2)$, the OLS estimator is not a consistent estimator for β when the trending component g_t is weak.

(3) *The regressor x_t has a strong trend*

In this case, we suppose that x_t has a strong trend that $T^{-d}\sum_{t=1}^{T}(g_t - \bar{g}_T)^2 \to S_g$ for some $d > 1$. The value of d is determined by the strength of the trend in x_t, i.e., stronger trends lead to larger values of d, and vice versa. Denote $\widehat{Q}^* = T^{-d}\sum_{t=1}^{T}(x_t - \bar{x})^2$ and $S_{xe}^* = T^{-d}\sum_{t=1}^{T}(x_t - \bar{x})(e_t - \bar{e})$. As T goes to infinity, $\widehat{Q}^*$ converges to S_g but S_{xe}^* converges to zero.

Therefore, for the OLS estimator $\widehat{\beta} = \beta + S_{xe}^*/\widehat{Q}^*$, we have

$$\widehat{\beta} \to \beta, \tag{2.50}$$

almost surely when sample size T tends to infinity, suggesting that $\widehat{\beta}$ is a consistent estimator for β as the bias term $S_{xe}^*/\widehat{Q}^*$ diminishes to zero with T.

To summarize, the problem of endogeneity induces bias in the OLS estimator when the sample size is small. For purely stationary regressors, the bias does not vanish with a growing sample size. When a weak trend is present in the regressor, the bias term is smaller than that in the stationary case by comparing the limits of the biases in equations (2.48) and (2.49). In both cases, the OLS estimator $\widehat{\beta}$ does not converge to the real value of β for large samples as well. When the trend is strong, the trending information dominates the correlation between the regressor and the regression error, reducing the bias in the OLS estimator to zero when the sample size tends to infinity. Therefore, the OLS estimator is consistent when the trend dominates the stationary errors.

For regressions with nonstationary regressors, Phillips and Hansen (1990) studied the endogeneity in the regression where the regressors follow pure random walk processes. For their regression model, we have $d = 2$ and the OLS estimator is consistent in large samples. However, the limit distribution of $T(\widehat{\beta}-\beta)$ is not centered around zero because of endogeneity. In other words, when the time series are $I(1)$, the consistency of the OLS estimator is not affected by endogeneity as the stochastic trend is relatively strong and dominates the time series. However, since the endogenous correlation causes a bias in the limit distribution of $T(\widehat{\beta}-\beta)$, the inference of β is somehow affected.

In the literature, the instrumental variable regression method is widely employed to establish consistent estimators of the coefficients in the regression models with endogeneity. In this book, we propose two different estimation methods to deal with the endogeneity issue in the weak trend regression when $d = 1$ and the strong trend regression when $d > 1$.

2.7 Trending Time Series in Predictive Regressions

Predictive regressions have been extensively studied in recent decades. In particular, y_{t+1} (for example, the financial stock returns) are regressed on the lagged value of certain predictive variable x_t (for example, the dividend-price ratio) in the form of

$$y_{t+1} = \alpha + \beta x_t + e_{t+1}, \tag{2.51}$$

where e_t is supposed to be stationary regression errors. The significance of β indicates the statistical evidence of the predictive power of x_t on the future values of y_t. Therefore, testing for the null hypothesis $\mathbb{H}_0 : \beta = 0$ is the essential goal of interest.

The economic and financial foundation for many predictive regressions in the literature is the Campbell–Shiller identity proposed by Campbell and Shiller (1988). They start from the formula of stock return that $r_{t+1} = \log((P_{t+1} + D_{t+1})/P_t)$, where P_t and D_t are the stock price and dividend paid at time t. Simple transformation of r_{t+1} gives

$$r_{t+1} = \Delta d_{t+1} + dp_t + \log(1 + e^{-dp_{t+1}}), \tag{2.52}$$

where $\Delta d_{t+1} = \log D_{t+1} - \log D_t$ is the log dividend growth and $dp_t = \log(D_t/P_t)$ is the log dividend-price ratio. The first-order Taylor expansion of the last term in equation (2.52) at the long-run mean of dp_t gives

$$r_{t+1} = k + \Delta d_{t+1} - \zeta dp_{t+1} + dp_t, \tag{2.53}$$

and k and ζ are constant parameters. Iterating equation (2.53) into the infinity future and assuming a standard no-bubble condition, Campbell and Shiller (1988) obtain the present-value identity

$$dp_t = -\frac{k}{1-\zeta} + E_t \sum_{j=1}^{\infty} \zeta^{j-1} r_{t+j} - E_t \sum_{j=1}^{\infty} \zeta^{j-1} \Delta d_{t+j}, \tag{2.54}$$

where E_t denotes the expectation conditional on time t. This identity shows that the log dividend-price ratio is approximated as the present value of future stock returns r_{t+j} and future dividend growth Δd_{t+j} for $j \geq 1$. As a result, low dp_t will predict a low future value of r_{t+j} and a high future value of Δd_{t+j} for the market as a whole. Hence, researchers have been seeking econometric evidence of the financial theory of the Campbell–Shiller present value identity using predictive regressions. The main question is can dp_t predict future values of r_t and Δd_t? Testing for predictability in equation (2.51) seems to be a simple practice of the conventional t-test for the significance of a single regression coefficient. However, some serious econometric issues may hinder correct inferences of β.

The first problem is that the error terms e_t in equations (2.51) and v_t in (2.58) are likely to be contemporaneously correlated. This induces the problem of "embedded endogeneity" as first proposed by Stambaugh (1999). Specifically, the two error terms are assumed to follow a joint normal distribution with mean zero and covariance matrix:

$$(e_t, v_t)' \sim \mathcal{N}(0, \Omega), \quad \text{for } \Omega = \begin{pmatrix} \sigma_e^2 & \sigma_{ev} \\ \sigma_{ev} & \sigma_v^2 \end{pmatrix}, \tag{2.55}$$

where $\sigma_{ev} \neq 0$. When $|\rho| < 1$, Stambaugh (1999) showed that $\mathrm{E}(\widehat{\beta} - \beta) = (\sigma_{ev}/\sigma_v^2)\mathrm{E}(\widehat{\rho} - \rho)$, where $\widehat{\beta}$ and $\widehat{\rho}$ are the OLS estimators in regression equations (2.51) and (2.58), respectively. This indicates that the OLS estimator of the predictive regression is biased in finite sample. Furthermore, Stambaugh (1999) approximated the bias in $\widehat{\rho}$ and therefore derives the bias in $\widehat{\beta}$ as

$$\mathrm{E}(\widehat{\beta} - \beta) = -\frac{\sigma_{ev}}{\sigma_v^2}\left(\frac{1+3\rho}{T}\right) + O(T^{-2}), \tag{2.56}$$

where the bias shrinks with sample size T. Namely, as $T \to \infty$, the Stambaugh bias diminishes and the OLS estimator of $\widehat{\beta}$ is consistent in large samples. Nevertheless, the bias in $\widehat{\beta}$ may cause misleading inference results when the sample size T is not sufficiently large.

To solve this problem, Amihud and Hurvich (2004) proposed a reduced-bias estimation method that first estimates equation (2.58) by the OLS method to obtain $\widehat{\rho}$ and then constructs the bias-corrected estimator $\widehat{\rho}^c = \widehat{\rho} + (1 + 3\widehat{\rho})/T + 3(1 + 3\widehat{\rho})/T^2$ and obtains the residuals by $\widehat{v}_t^c = x_t - \widehat{\mu}^c - \widehat{\rho}^c x_{t-1}$ for $t = 2, 3, \ldots, T$. The reduced-bias estimator $\widehat{\beta}^c$ is the coefficient of x_t in the extended OLS regression

$$y_{t+1} = \alpha + \beta x_t + \phi \widehat{v}_{t+1}^c + u_{t+1}, \tag{2.57}$$

for $t = 1, 2, \ldots, T - 1$. There are some follow-up papers based on the work of Amihud and Hurvich (2004). Amihud *et al.* (2009) proposed a new test for multipredictor regressions in which x_t follows a stationary VAR(1) model, and Amihud *et al.* (2010) investigated the AR(p) case that both y_{t+1} and x_{t+1} are regressed on $x_t, \ldots, x_{t-p+1}$ rather than the AR(1) model in equations (2.51) and (2.58). Instead of assuming x_t is stationary with $|\rho| < 1$ in the works of Stambaugh (1999); Amihud and Hurvich (2004); Amihud *et al.* (2009), Cai and Wang (2014) considered the case that x_t is nearly integrated with $\rho = 1 + c/T$ with $c \leq 0$ and proposed using a projection method by setting $e_t = \phi v_t + u_t$.

The second problem originates from the mismatching persistence of the time series of y_t and x_t on the two sides of the regression equation (2.51). In the empirical analysis, the regressand y_t (for example, the excess stock returns in Figure 2.1) is stationary exhibiting very low autocorrelation and even behaves like a WN process. On the other hand, the predictors (for instance, the dividend-price ratio dp_t, the earning price ratio ep_t, and the book-to-market ratio bm_t) are found to be highly persistent and exhibit trend-like behaviors, as shown in Figure 2.1. In many research papers, they are often assumed

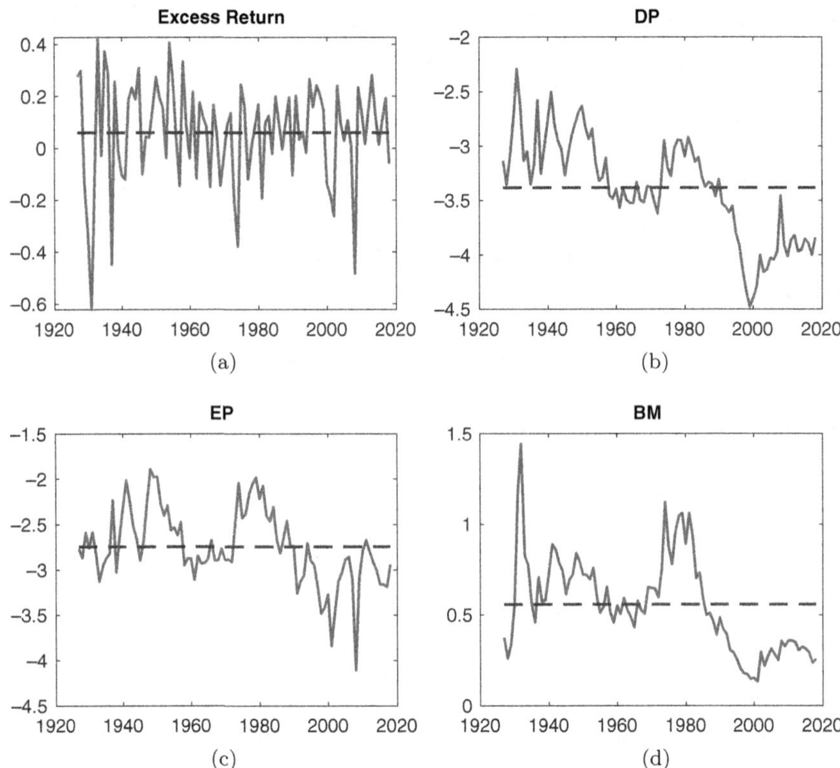

Figure 2.1. Stock returns and financial ratio predictors dp_t, ep_t, and bm_t.

to follow an autoregressive process formulated as[3]

$$x_{t+1} = \mu + \rho x_t + v_{t+1}, \tag{2.58}$$

where the first-order autocorrelation ρ is found to be close to one. For example, Table 4 in the work of Campbell and Yogo (2006) gives the 95% confidence intervals for ρ at the annual, quarterly, and monthly frequencies with respect to dp_t and ep_t, and most of their lower limits are above 0.9.

Campbell and Yogo (2006) and Cai and Wang (2014) addressed the impact of high persistency in the predictors on econometric inferences for the predictive regressions. The autoregressive coefficient is

[3]Alternatively, they are also modeled as $x_{t+1} = \mu + v_{t+1}$, where $v_{t+1} = \rho v_t + u_{t+1}$ and $(e_t, u_t)'$ follows an $i.i.d.$ joint normal distribution.

modeled as $\rho = 1 + c/T$ in which c is a constant value, and asymptotic properties of the OLS estimator $\widehat{\beta}$ in equation (2.51) are also established in these papers. This allows x_t to be a unit root process ($c = 0$), a stationary but nearly integrated or local-to-unity process ($c < 0$), and even an explosive process ($c > 0$). Their asymptotic results and the corresponding statistical tests depend on the persistence level c as a nuisance parameter.

The two sides of equation (2.51) are not balanced in their persistence, however, we are predicting future values of a stationary and almost uncorrelated non-trending variable using a nonstationary highly persistent trending predictor. Consequently, the imbalance in the persistence of the predictive regression forces β to be rather small in magnitude if it is not zero, and it is also very likely to be statistically insignificant. On the other hand, the high persistence and even nonstationary in the regressor x_t make econometric modeling and coefficient testing much more difficult and complicated than the usual t-test for stationary regressions. This is the reason why predictability is rarely detected in empirical results using predictive regression (see the works of Campbell and Yogo, 2006; Welch and Goyal, 2008, and the references therein).

To solve this imbalanced issue in the predictive regressions, Ren *et al.* (2019) provided a general framework of balanced predictive regression. Specifically, for the case with only one predictor, they augment the conventional predictive regression model by adding one more lag as

$$y_{t+1} = \alpha + \beta_1 x_t + \beta_2 x_{t-1} + e_{t+1}, \tag{2.59}$$

where the predictor $x_{t+1} = \mu + \rho x_t + v_{t+1}$ is highly persistent with $\rho = 1 - c/T^\gamma$ for $c \geq 0$ and $\gamma \in (0, 1]$. The error terms e_t and v_t are assumed to be correlated. In fact, model (2.59) has the effect of using the approximate first difference of x_t as the predictor, and balance is achieved through the interaction (cointegration) between the two highly persistent lags. In the new model, predictability is tested by examining the null hypothesis $\mathbb{H}_0 : \beta_1 = \beta_2 = 0$. If the data is generated by equation (2.59), testing for predictability using equation (2.51) is equivalent to testing for $\mathbb{H}_0 : \beta_1 + \beta_2 = 0$ in model (2.59). It is possible that we could not reject $\mathbb{H}_0 : \beta_1 + \beta_2 = 0$ and find no predictability based on equation (2.51), but in fact the approximate difference may help predict y_t due to $\beta_1 \approx -\beta_2 \neq 0$. As a result, the predictability is missed based on the conventional predictive regression but captured by the augmented balanced model.

When multiple predictors are involved, their persistency or trending patterns may cancel out by cointegration producing a stationary sequence that is capable of predicting stock returns. Lettau and Ludvigson (2001) constructed the *cay* predictor for stock returns by assuming that consumption (c_t), asset holdings (a_t), and labor income ($\check{y}_t$) are cointegrated and share a common long-term trend. Specifically, the cointegration relationship is estimated by a dynamic least squares regression model:

$$c_t = \alpha + \beta_1 a_t + \beta_2 \check{y}_t + \sum_{i=-k}^{k} b_{a,i} \Delta a_{t-i} + \sum_{i=-k}^{k} b_{\check{y},i} \Delta \check{y}_{t-i} + \varepsilon_t, \quad (2.60)$$

where the leads and lags of the first difference of a_t and y_t are included in the regression to eliminate the effects caused by the problem of endogeneity. Then, the predictor $\widehat{cay}_t$ is obtained as the estimated trend deviation by $\widehat{cay}_t = c_t - \widehat{\beta}_1 a_t - \widehat{\beta}_2 \check{y}_t$, which displays significant predictability for the U.S. stock returns. Instead of first obtaining cointegration error cay_t and using it as a single predictor, Ren *et al.* (2019) suggested employing a more flexible approach by

$$r_{t+1} = \alpha + \beta_1 c_t + \beta_2 a_t + \beta_3 \check{y}_t + e_{t+1}, \quad (2.61)$$

where the persistent variables are directly included in the predictive regression model. This model is balanced due to the cointegration among c_t, a_t, and $\check{y}_t$. Testing the null hypothesis of no predictability $\mathbb{H}_0 : \beta_1 = \beta_2 = \beta_3 = 0$ against the alternative $\mathbb{H}_1 : \beta_1 \neq 0$ or $\beta_2 \neq 0$ or $\beta_3 \neq 0$ in model (2.61) is equivalent to testing the corresponding hypothesis concerning the coefficient of cay_t in the regression of r_{t+1} on cay_t.

Another approach to solving the imbalanced predictive regression is to transform the highly persistent predictors so as to reduce their persistence before we include them in predictive regressions. Lettau and Van Nieuwerburgh (2008) argued that there may exist shifts in the steady state mean of the economy and we have to adjust the financial ratio predictors for such shifts. Figure 2.2 is borrowed from the work of Lettau and Van Nieuwerburgh (2008); it shows that after

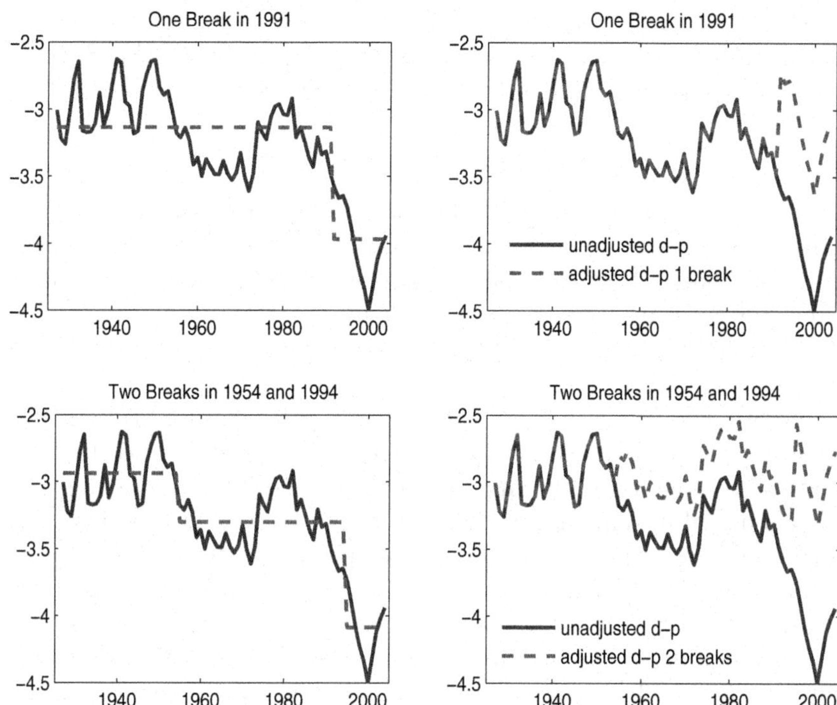

Figure 2.2. Adjustments for the shifts in the mean of the dividend-price ratio with one break in 1991 and two breaks in 1954 and 1994 (Figure 2 in the work of Lettau and Van Nieuwerburgh (2008)).

adjusting for one break in 1991 or two breaks in 1954 and 1994, the mean-adjusted dividend-price ratio dp_t turns out to behave like a stationary process. Using the adjusted dividend-price ratios instead of the raw series as the predictor variable, Lettau and Van Nieuwerburgh (2008) found that the adjusted dividend-price ratio significantly predicts stock returns with relatively stable coefficients. When using the original series of dp_t, however, the estimated coefficients are insignificant and not stable on each of the subsamples.

The mean shifts in the work of Lettau and Van Nieuwerburgh (2008) are estimated and tested based on the time series of dp_t itself as a statistical fact without providing further explorations of possible fundamental reasons why such structural breaks occur. The researcher also needs a relatively long range of samples available to detect and estimate the time points of structural breaks.

Favero *et al.* (2011) linked the dividend-price ratio with a natural demographic variable, the middle-aged to young ratio MY_t. Specifically, they establish an extended predictive regression model as

$$r_{t+1} = \alpha + \beta_1 p_t + \beta_2 d_t + \beta_3 MY_t + e_{t+1}, \tag{2.62}$$

where $(p_t, d_t, MY_t)'$ are expected to be cointegrated, and their combination gives a stationary process, which they regard as the information component that drives long-horizon stock market fluctuations after the noise in short-horizon stock market fluctuations dies out. Results show that at the 5% significance level, β_1, β_2, and β_3 are significant and the null hypothesis of $\mathbb{H}_0 : \beta_1 = -\beta_2$ is rejected. Therefore, the predictivity of the dividend-price ratio is statistically evident, however, the cointegrating vector for $(d_t, p_t)'$ is not $(1, -1)$ that the trends in the two time series cannot cancel out.[4] Meanwhile, MY_t is useful in removing the slowly evolving trend in the dividend-price ratio. It improves the predictive power of the dividend-price ratio for stock returns.

Recently, Yu *et al.* (2023) also argued that $dp_t = d_t - p_t$ is not stationary because the trends in d_t and p_t may not cancel out[5] perfectly with some trending component remaining in dp_t. Therefore, they employed a data-driven nonparametric method to decompose the dividend-price ratio into a slow-moving component f_t that reflects the time-varying local mean as the remaining trend, and a cyclical component c_t that reflects the transitory deviations of the ratio from its local mean. Specifically, a nonstationary trending predictor is separated as

$$x_t = f_t + c_t \tag{2.63}$$

in an additive manner. The empirical results show that the cyclical component c_t delivers substantially improved forecast gains of stock returns and dividend growth relative to the original financial ratio x_t and the historical average benchmark. Conversely, the slow-moving component f_t fails to predict returns, and therefore they are found to

[4]Here, we consider a rearranged model $r_{t+1} = \alpha - \beta_1 dp_t + (\beta_1 + \beta_2)d_t + \beta_3 MY_t + e_{t+1}$.

[5]Note that $p_t = \log P_t$ and $d_t = \log D_t$, where P_t is the stock price and D_t is the dividend paid at time t.

disguise the predictive information contained in x_t for stock returns and cash flows.

2.8 Aggregate Income and Consumption

For the empirical research question when there exist both issues of nonstationarity and endogeneity, an old popular example is the linear regression of personal consumption against income at the aggregate level. The regression model is simply formulated as

$$C_t = \alpha + \beta I_t + e_t, \tag{2.64}$$

for $t = 1, 2, \ldots, T$, where C_t and I_t represent the logarithms of aggregate personal consumption expenditure and aggregate personal disposable income. The regression coefficient β is expected to reflect the long-run marginal propensity to consume (MPC), which measures the proportion to consume when a typical consumer earns an additional one percent of his or her income. The error term e_t, however, is assumed to be stationary implying that the trends in C_t and I_t cancel out regardless of their exact forms. Figure 2.3 plots the logarithms of the U.S. aggregate personal disposable income and aggregate personal consumption expenditure from 1959Q1 to 2019Q4. The figure shows that both series exhibit upward trends instead of fluctuating around a stable level, thus they are nonstationary trending over time.

The regression of aggregate income over consumption has been extensively studied as this simple model is related to the verification of the PIH, which has been a very popular topic since the 1970s (Campbell, 1987). However, researchers haven't reached a consensus on the exact nature of trend in both series of consumption and income in the literature. For example, Hall (1978) solved the consumer's optimization problem under the condition of rational expectations. He concluded that consumption should follow a random walk process and that the changes are not predictable under the PIH, given that the real interest rate is a constant value over time. In other words, consumption tracks the permanent income, and it is not sensitive to the changes in current income.

However, opposite conclusions are found in some other papers. Flavin (1981) developed a structural econometric model of consumption to estimate the excess sensitivity of consumption to current

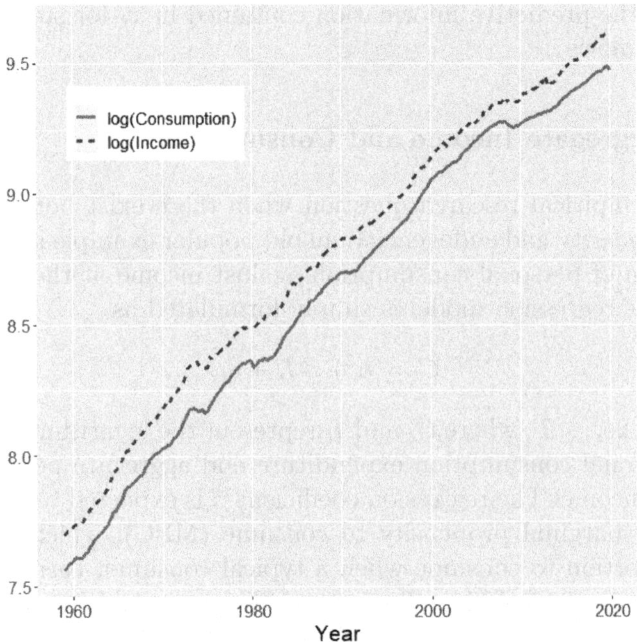

Figure 2.3. Logarithms of the quarterly U.S. aggregate personal income and consumption from 1959Q1 to 2019Q4.

income. Such excess sensitivity should be zero under the PIH. The empirical result shows a strong effect of excess sensitivity in the consumption to current income. Therefore, the PIH is rejected. Similar results are also found in the works of Flavin (1984) and Bernanke (1985). In these papers, they assume that the time series of income is a stationary process around a deterministic trend.

With the prevalence and application of unit root tests, Mankiw and Shapiro (1985) conducted their analysis based on the statistical evidence in the work of Nelson and Plosser (1982) that income and consumption exhibit unit roots. They show that excess sensitivity is favorable if we ignore the unit-root structure in the trending data and conduct inappropriate detrending, which could bring spurious cycles in the transformed data (also see the work of Nelson and Kang (1981, 1984)). Therefore, Flavin's conclusions are likely to be biased and not reliable. King *et al.* (1991) proposed a cointegration model with the cointegrating vector fixed as $(1, -1)$ for the logarithms of consumption and income, and the model is regarded as a special version of the PIH with a hundred percent MPC. Han and Ogaki (1997) established the cointegration relationship between both the stochastic trend and the deterministic trend, and they find that both trends are cointegrated, implying that the post-war U.S. saving rate is stable in the long run.

Researchers are aware that the estimation of such a regression relationship of consumption over income also suffers from the problem of endogeneity as both of them belong to a system of simultaneous equations. At the aggregate level for a closed economy, total income I_t originates from total consumption:

$$I_t = C_t + D_t, \tag{2.65}$$

where C_t is the aggregate consumption expenditures and D_t is the aggregate nonconsumption expenditure. Hence, by combining equations (2.64) and (2.65), we find that aggregate income and consumption determine each other simultaneously. It is then straightforward to show that $\text{Cov}(I_t, e_t) = \sigma_e^2/(1-\beta) \neq 0$ when $\beta \neq 1$. Therefore, the regressor I_t and the error term e_t are correlated in equation (2.64) so that there exists the problem of endogeneity if we only consider and estimate equation (2.64). Phillips and Hansen (1990) developed the bias-correction method for the cointegration models with endogeneity. In the work of Hansen and Phillips (1990), the authors applied the bias-correction method to the cointegration model of the per capita personal consumption over the per capita personal income. In their paper, the PIH is not rejected as the unit coefficient is included in the 95% confidence interval of the estimated coefficient.

To summarize, in the regression of aggregate income over aggregate consumption, both the dependent and explanatory variables are nonstationary with upward trends. Therefore, conventional estimation and inference methods for the linear regression models cannot be applied as the critical assumptions (stationary and exogeneity conditions) are not satisfied. Moreover, how the empirical time series are trended is not exactly known. Meanwhile, the regressor and the error term are likely to be correlated due to simultaneous causality. The OLS estimate of the long-run MPC is likely to be biased and inconsistent, and the usual t-test may lead to incorrect inferential conclusions.

2.9 Summary

In recent decades, there has been significant development and application of trending time series models, incorporating a diverse range of deterministic time trends, stochastic trends, or a combination of both. They are designed to directly analyze trending time series and their regression relationships. This approach circumvents the removal of trends, which can be problematic, even though differenced or filtered counterparts are more likely to exhibit stationarity and are thus easier to model within the conventional framework. By directly engaging with trending time series, these models offer a more nuanced and comprehensive understanding of the underlying dynamics, avoiding potential information loss associated with detrending methods.

The concept of cointegration, widely employed in these models, illustrates common stochastic trends indicating long-run equilibrium relationships among crucial economic variables, such as income and consumption. Linear cointegration has been a popular focus, showcasing relationships between variables that persist over the long term. Additionally, there has been a growing interest in exploring nonlinear cointegration and fractional cointegration models, which may take parametric or nonparametric forms. These advancements reflect a nuanced understanding of the complexities inherent in economic time series data.

An important challenge in estimating and inferring regression relationships between trending time series is the problem of endogeneity. The presence of endogeneity can impede accurate estimation and inference, necessitating adjustments to account for the inherent relationship between regressors and regression errors. Addressing the problem of endogeneity is crucial for ensuring the reliability and validity of the results obtained from trending time series models.

2.10 Bibliographical Notes

For the time-varying coefficient regression model $y_t = x_t'\beta_t + e_t$, in addition to being deterministically time-varying as $\beta_t = \beta(\tau_t)$, the coefficient vector can also be stochastically time-varying. This type of model is known as a random coefficient model. The first case is the stationary case. That is, β_t is stationary over time and can be estimated by the GLS method, the maximum likelihood estimation (MLE) method, or the Bayesian method (Swamy and Tinsley, 1980); (Harvey and Phillips, 1982); (Kim and Nelson, 1999). The second case is that β_t is nonstationary. For example, β_t has a unit root $\beta_t = \beta_{t-1} + v_t$ as studied in the works of Rosenberg (1973), Cooley and Prescott (1976), Nicholls and Pagan (1985), Giraitis *et al.* (2014), and Hastie and Tibshirani (2018). Such models are widely applied in practice, for instance, in the works of Stock and Watson (1996), Brown *et al.* (1997), Cogley and Sargent (2005), Kim and Nelson (2006), Cogley *et al.* (2010), and Dangl and Halling (2012). Fu *et al.* (2023) proposed two tests to distinguish these three specifications, i.e., the stochastically stationary case, the stochastically nonstationary case, and the deterministic function of time case for β_t.

Chapter 3

Time Series Regressions with Weak Trends

> Whenever possible, we recommend attempting to relate apparent trends to appropriate underlying phenomena, whether economic, demographic, political, legal, technological, or physical.
>
> — Granger (2011)

In a typical trending regression model, a time series is regressed against one or more other trending time series. The strength of a trend varies depending on its growth rate over time. Specifically, a stronger trend tends to grow faster toward infinity compared to a relatively weaker one. In particular, a *strong trend* denotes the category of trends that expand to infinity over time, while a *weak trend* is bounded that does not diverge to infinity and behaves like a slowly and smoothly moving mean.

Meanwhile, the regression relationships between trending time series are often affected by the problem of endogeneity, which is a common issue in economic analysis. Namely, if the regressors are correlated with the regression errors, the estimation of the regression coefficients may be biased and inconsistent, making inferences on the actual values of coefficients unreliable. This can significantly impact the accuracy of the results of the regression analysis.

The actual impacts, however, depend on the strengths of trends in the regressors. In general, the presence of a strong trend in the regressor alleviates the biasedness and inconsistency in the OLS estimators as the strong trending information usually plays a dominant role over the stationary components in the trending time series. On the contrary, the OLS estimators are biased and inconsistent when the regressors contain weak trends, which are components that do not dominate the total variation in the time series.

This chapter investigates the linear trending regression model in which the regressors contain weak trends. At the same time, we allow for the existence of the problem of endogeneity in the regression models. To consistently estimate the coefficients in the model, we propose a nonparametric control function approach and provide asymptotic properties of the proposed estimators. Simulated and empirical examples are provided to demonstrate the implementation steps of the proposed method.

3.1 Model Specifications

For a nonstationary trending time series, whether its time trend is weak or strong depends on the trending strength or magnitude. Formally, we let $\sum_{t=1}^{T} g_t^2 = O_p(T^d)$ and define g_t as a weak trend when $d = 1$. Otherwise, when $d > 1$, we say g_t has a strong trend. In this chapter, we consider the linear regression model in which the regressors have weak trends. The model is formulated as

$$y_t = \alpha + x_t'\beta + e_t, \tag{3.1}$$

$$x_t = g(\tau_t) + v_t, \tag{3.2}$$

for $t = 1, 2, ..., T$, where $\tau_t = t/T$, $x_t = (x_{1t}, x_{2t}, \ldots, x_{kt})'$ is a $k \times 1$ vector of trending regressors, $\beta = (\beta_1, \beta_2, \ldots, \beta_k)'$ is a $k \times 1$ vector of unknown coefficients, and α is the intercept term. The weak trend functions $g(\tau) = (g_1(\tau), g_2(\tau), \ldots, g_k(\tau))'$ are continuous and second-order differentiable on $[0, 1]$. They are also integrable in mean and variance that $\lim_{T\to\infty} \bar{g}_T = \int_0^1 g(\tau)d\tau \triangleq \bar{g} < \infty$, where $\bar{g}_T = T^{-1}\sum_{t=1}^{T} g(\tau_t)$ and $\lim_{T\to\infty} T^{-1}\sum_{t=1}^{T}(g(\tau_t) - \bar{g}_T)(g(\tau_t) - \bar{g}_T)' = \int_0^1 (g(\tau) - \bar{g})(g(\tau) - \bar{g})'d\tau \triangleq \Sigma_g$, where Σ_g is a positive definite matrix. The error terms e_t and v_t are stationary processes with $\mathrm{E}(e_t) = 0$, $\mathrm{E}(v_t) = 0$, $\mathrm{E}(e_t^2) = \sigma_e^2 < \infty$, and $\mathrm{E}(v_t v_t') = \Sigma_v$ is a positive definite matrix. As a result, when the trend is weak, we have $\lim_{T\to\infty} T^{-1}\sum_{t=1}^{T} x_t x_t' = \Sigma_g + \Sigma_v \triangleq \Sigma_x$, which is also positive definite.

Note that the weak trend restriction rules out the common case of the usual linear time trends. Specifically,

$$x_t = a_1 + b_1 t + v_t, \tag{3.3}$$

$$x_t = a_2 + b_2(\tau_t) + v_t \tag{3.4}$$

are two different kinds of trends, for $t = 1, 2, \ldots, T$. In equation (3.3), the linear trend is strong because $\sum_{t=1}^{T} t^2 = O(T^3)$ in which $d = 3$. While in equation (3.4), x_t has a weak trend because $\sum_{t=1}^{T} (\tau_t)^2 = O(T)$ in which $d = 1$. When sample size T gets larger, the strong linear trend diverges to infinity as long as $b_1 \neq 0$, while the weak linear trend retains its shape but becomes denser on [0, 1]. Therefore, when sample size T is fixed, the two forms have no essential difference because $b_1 = b_2/T$. However, when $T \to \infty$, if the true model is equation (3.4), then using model (3.3) is wrong because $b_1 \to 0$ with T. If the true model is equation (3.3), then using model (3.4) is incorrect because the slope coefficient $b_2 \to \infty$ with T. A simple approach to differentiate models (3.3) and (3.4) is to see whether x_t exhibits an explosive manner with T according to certain economic or financial theories.

We assume that the stationary error terms e_t and v_t in equations (3.1) and (3.2) are correlated with an unknown structure. Such correlation brings about the correlation between x_t and e_t and causes the problem of endogeneity in the regression model of equation (3.1). We have shown in the previous chapter that in such regression models with endogeneity, the OLS estimator is biased and inconsistent. The consequence is that it does not converge to the true values of α and β even when the sample size T tends to infinity.

To solve this problem and to obtain consistent estimations of α and β, we employ a *nonparametric control function approach.* We assume that the error terms e_t and v_t satisfy a general nonlinear relationship:

$$e_t = \lambda(v_t) + u_t, \tag{3.5}$$

for $t = 1, 2, \ldots, T$, where $\lambda(v_t) = \mathrm{E}(e_t|v_t)$ is an unknown smooth function defined on $\mathbb{R}^k \to \mathbb{R}^1$, representing the conditional mean of e_t given the value of v_t, and $u_t = e_t - \mathrm{E}(e_t|v_t)$ denotes the remainder. Note that by the *law of iterated expectations*, both $\lambda(v_t)$ and u_t have zero mean that $\mathrm{E}(\lambda(v_t)) = \mathrm{E}(\mathrm{E}(e_t|v_t)) = \mathrm{E}(e_t) = 0$

and $\mathrm{E}(u_t) = \mathrm{E}(e_t - \mathrm{E}(e_t|v_t)) = 0$. Equation (3.5) could be viewed as a decomposition of e_t using its conditional mean on v_t. A similar method is also used by Amihud and Hurvich (2004) and Cai and Wang (2014), in which the authors assume a linear relationship between e_t and v_t that $e_t = v_t'\delta + u_t$, and they call it as a *projection method*, which extends the linear regression model to

$$y_t = \alpha + x_t'\beta + v_t'\delta + \tilde{u}_t, \tag{3.6}$$

where the problem of endogeneity is resolved if the new error term $\tilde{u}_t$ is uncorrelated with x_t and v_t. Nonetheless, if $\tilde{u}_t$ is not orthogonal to v due to the existence of higher-order terms of v in $\lambda(v)$, for example, $e_t = v + v^2 + v^3 + u_t$, then a linear projection may not be sufficient to obtain an efficient estimator for the unknown coefficients.

The exact functional forms of $g(\tau)$ and $\lambda(v)$ are usually unknown. Pre-specified forms on them may be highly misspecified and cause incorrect estimation and inference results. Therefore, we do not impose certain parametric forms on them and leave them as nonparametric functions of τ and v. The advantage is that the nonparametric functions are estimated in a data-driven manner without relying on parametric structures. In this way, we let the data "speak for themselves" and can avoid the risk of model misspecification. The disadvantage is the potential loss of efficiency that estimating nonparametric models often requires a much longer length of sample compared with parametric models.

After decomposing the error term in model (3.1), we substitute the regression error e_t by its expression in equation (3.5) and extend the original linear regression model (3.1) to a semiparametric partially linear model formulated as

$$y_t = \alpha + x_t'\beta + \lambda(v_t) + u_t, \tag{3.7}$$

for $t = 1, 2, \ldots, T$. According to equation (3.5), the new error term is defined as $u_t = e_t - \mathrm{E}(e_t|v_t)$, and we have $\mathrm{E}(u_t v_t) = \mathrm{E}[(e_t - \mathrm{E}(e_t|v_t))v_t] = 0$ so that the new error term u_t in the partially linear model is uncorrelated with both x_t and v_t. Hence, the problem of endogeneity disappears in the extended model of equation (3.7).

The semiparametric partially linear model of equation (3.7) has been extensively studied and widely applied in the literature. The estimation method as well as the corresponding asymptotic properties are well addressed in the works of Robinson (1988), Härdle *et al.* (2000), Gao (2007), and Li and Racine (2007), as well as many other related papers. A widely accepted necessary condition to ensure the identifiability of the model is that the matrix

$$\Sigma = \mathrm{E}[(x_t - \mathrm{E}(x_t|v_t))(x_t - \mathrm{E}(x_t|v_t)'] \tag{3.8}$$

must be positive definite. Otherwise, the coefficient vector β is not estimable.

Our model in equation (3.7) is composed of a linear parametric part $\alpha + x_t'\beta$ and a possibly nonlinear nonparametric part $\lambda(v_t)$. However, they differ from the usual partially linear models in the literature in two major aspects. First, the variables in the parametric component x_t are nonstationary. While in most of the existing research papers, x_t is assumed to be stationary. The variable vector v_t in the nonparametric component is not observed and must be estimated based on equation (3.2), i.e., the data generating mechanism of x_t. While in the literature, v_t is usually observable. Second, the regressor in the linear components x_t and v_t differs only by a deterministic trend $g(\tau_t)$, which implies that $\mathrm{E}[x_t|v_t] = x_t$. It is worth noting that there is a potential identification problem for β. Specifically, the identification matrix $\mathrm{E}[(x_t - \mathrm{E}(x_t|v_t))(x_t - \mathrm{E}(x_t|v_t)']$ degenerates to zero as $g(\tau_t)$ is deterministic and $\mathrm{E}[x_t|v_t] = x_t$.

3.2 The Estimation Method

The nonparametric regression model has much flexibility to capture the functional relationship between the dependent variable and the regressors without being incorrectly restricted to certain parametric specifications. The estimation of a nonparametric regression model is purely data-driven. Specifically, we consider a general nonparametric regression model of the form

$$x_t = m(z_t) + e_t, \tag{3.9}$$

for $t = 1, 2, \ldots, T$, where the nonparametric function $m(z_t) = \mathrm{E}(x_t|z_t)$ representing the conditional mean of x_t given z_t, and it can also be regarded as the regression function of x_t over z_t. The error term $e_t = x_t - \mathrm{E}(x_t|z_t)$ is an ergodic stationary process with zero mean and finite variance.

We denote by $\widehat{\mathrm{E}}_h(x_t|z_t)$ the nonparametric kernel estimator of $\mathrm{E}(x_t|z_t)$, for example, the nonparametric local constant estimator or the local polynomial estimator as proposed and discussed in the works of Cai *et al.* (2000); Fan and Gijbels (1996); Härdle *et al.* (2000); Li and Racine (2007), where h denotes the bandwidth selected as a tuning parameter to adjust for the smoothness of the estimated function for $m(z_t)$. In particular, the local constant estimator (also called the Nadaraya–Watson estimator) is given by

$$\widehat{\mathrm{E}}_h(x_t|z_t = z) = \sum_{s=1}^{T} w(z_s, z)x_s, \tag{3.10}$$

where $w(z_s, z) = K_h(z_s - z)/\sum_{l=1}^{T} K_h(z_l - z)$, for $s = 1, 2, \ldots, T$, and $K_h(u) = K(u/h)/h$ with bandwidth h, and $K(\cdot)$ is a properly defined kernel function.

The regression equation (3.2) is similar to (3.9), where $m(z_t)$ is replaced by $g(\tau_t)$, the nonparametric time trend of the time series sequence x_t. Instead of being a random variable, $\tau_t = t/T$ is an equally spaced fixed-design sequence on [0, 1]. By replacing z_s with τ_s, the local constant kernel estimator for $m(x_t|z_t)$ becomes the nonparametric estimator of trend in x_t:

$$\widehat{g}(\tau) = \sum_{s=1}^{T} w_s(t)x_s, \tag{3.11}$$

where $w_s(t) = K_{st}/\sum_{l=1}^{T} K_{lt}$ for $s = 1, 2, \ldots, T$, in which $K_{st} = K_h(\tau_s - \tau_t)$.

The standard estimation procedures for the semiparametric partially linear model of equation (3.7) are proposed by Robinson (1988) and extensively discussed by Gao (2007). The key of the estimation method is to eliminate the nonparametric term $\lambda(v_t)$ and transform the model to a purely parametric one. Specifically, we take the conditional expectations on both sides of equation (3.7) with respect to v_t and obtain

$$\mathrm{E}(y_t|v_t) = \alpha + \mathrm{E}(x_t|v_t)'\beta + \lambda(v_t), \tag{3.12}$$

for $t = 1, 2, \ldots, T$, where we utilize the condition that $\mathrm{E}(u_t|v_t) = 0$. Subtracting from equation (3.7), we are able to eliminate the nonparametric term $\lambda(v_t)$ and the intercept term α with

$$\breve{y}_t = \breve{x}_t'\beta + u_t, \tag{3.13}$$

for $t = 1, 2, \ldots, T$, where $\breve{y}_t = y_t - \mathrm{E}(y_t|v_t)$ and $\breve{x}_t = x_t - \mathrm{E}(x_t|v_t)$. The conditional mean $\mathrm{E}(y_t|v_t)$ and $\mathrm{E}(x_t|v_t)$ could be estimated by the nonparametric estimation method, for example, the estimator in equation (3.10). Then, the regression coefficient vector β in equation (3.13) could be estimated by least squares method:

$$\breve{\beta} = \left(\sum_{t=1}^{T} \breve{x}_t \breve{x}_t' \right)^{-1} \sum_{t=1}^{T} \breve{x}_t \breve{y}_t. \tag{3.14}$$

The application of this estimator, however, encounters several obstacles for our model in practice. First, the variable v_t is not directly observed so $\breve{x}_t$ and $\breve{y}_t$ are not computable. To solve this problem, we have to estimate the trend functions $g(\tau)$ in (3.2) based on the observed sequence of x_t, and then obtain $\widehat{v}_t$ as the corresponding residual sequence. The nonparametric estimator $\widehat{g}(\tau_t)$ is expected to be an appropriate estimator of the real trend as precise as possible, for example, the kernel estimator in (3.11). Then, we could replace v_t by $\widehat{v}_t = x_t - \widehat{g}(\tau_t)$ when we compute the conditional expectations $\mathrm{E}(y_t|v_t)$ and $\mathrm{E}(x_t|v_t)$.

Second, the usual identification condition for the semiparametric partially linear model is not satisfied. Namely, Σ in (3.8) degenerates to a zero matrix when $g(\tau_t)$ is a deterministic trend function in equation (3.2) as the data-generating process of x_t. Note that this identification condition is imposed on the population distributions as in equation (3.8). Fortunately, we find that the sample analog $T^{-1} \sum_{t=1}^{T} \breve{x}_t \breve{x}_t'$ may not necessarily reduce to 0 if the trends satisfy the condition that Σ_g is a positive definite matrix. Therefore, the conventional estimation method is still applicable.

The rationale behind this is that the probability limit of the sample formula $T^{-1} \sum_{t=1}^{T} \breve{x}_t \breve{x}_t'$ is not $\mathrm{E}(\breve{x}_t \breve{x}_t') = 0$ when x_t follows equation (3.2). Instead, it converges to Σ_g. The main reason is that although $\mathrm{E}(x_t|v_t) = \mathrm{E}(g(\tau_t) + v_t|v_t) = x_t$, the nonparametric estimator $\widehat{\mathrm{E}}_h(x_t|v_t) - v_t$ converges to $\bar{g}$, where $\bar{g} = \int_0^1 g(\tau) d\tau$. Intuitively, as the nonparametric kernel estimator for conditional mean, $\widehat{\mathrm{E}}(x_t|v_t)$ is an operator for averaging x_t given the value of v_t. The value of the random variables x_t and v_t only differ by a deterministic trend term $g(\tau_t)$, which has no correlation with the random value of v_t. Therefore, when given the value of v_t, the estimated value of $\widehat{\mathrm{E}}(x_t|v_t)$ is $\bar{g} + v_t$. Figure 3.1 shows the scatter plot of x_t versus v_t, where v_t is

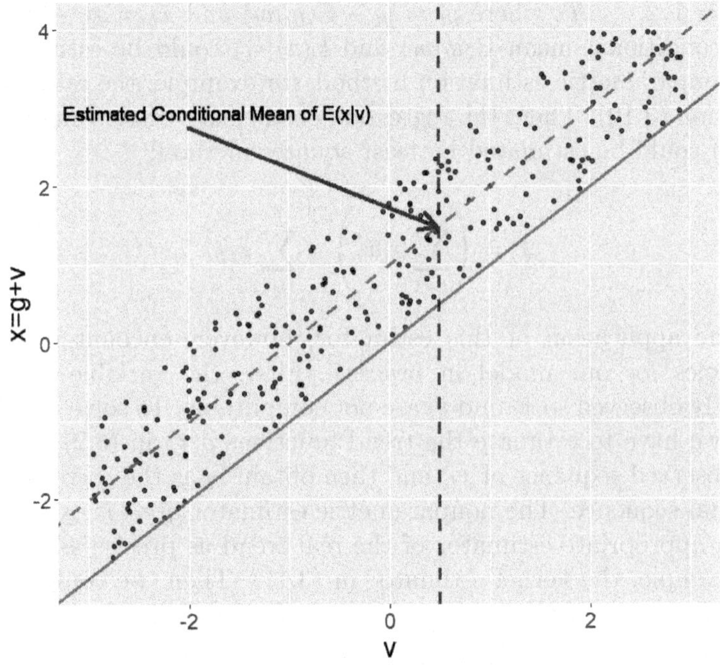

Figure 3.1. Conditional mean of x_t given v_t.

generated independently from a uniform distribution of $[-3, 3]$ and $x_t = 2t/300 + v_t$, for $t = 1, 2, \ldots, 300$. Therefore, $\bar{g} = 1$. The two 45° lines show the linear relationship of $x_t = v_t$ (solid line) and $x_t = \bar{g}+v_t$ (dashed line). Hence, given $v_t = 0.5$, the estimated conditional mean of x_t using the kernel estimation method is 1.5.

Therefore, a feasible estimator for β is

$$\widehat{\beta} = \left(\sum_{t=1}^{T} \widehat{x}_t \widehat{x}_t'\right)^{-1} \sum_{t=1}^{T} \widehat{x}_t \widehat{y}_t, \tag{3.15}$$

where $\widehat{x}_t = x_t - \widehat{E}_h(x_t|\widehat{v}_t)$ and $\widehat{y}_t = y_t - \widehat{E}_h(y_t|\widehat{v}_t)$. Then, the intercept term could be estimated by

$$\widehat{\alpha} = \frac{1}{T}\sum_{t=1}^{T}(y_t - x_t'\widehat{\beta}). \tag{3.16}$$

Finally, the endogenous relationship between e_t and v_t could be estimated by the nonparametric regression

$$\widehat{\lambda}(v) = \widehat{E}_h(\widehat{e}_t|\widehat{v}_t = v), \tag{3.17}$$

where $\widehat{e}_t = y_t - \widehat{\alpha} - x_t'\widehat{\beta}$.

3.3 Assumptions and Asymptotic Results

Before we formally establish the asymptotic results, we first introduce some useful assumptions.

Assumption 3.3.1. We assume that $g(\tau) = (g_1(\tau), \ldots, g_k(\tau))'$ is a $k \times 1$ vector of functions and each $g_i(\cdot)$ is a continuous and second-order differentiable function defined on [0,1] for $i = 1, 2, \ldots, k$. We also assume that as $T \to \infty$, the probability limit

$$\bar{g}_T = \frac{1}{T}\sum_{t=1}^{T} g(\tau_t) \longrightarrow \int_0^1 g(\tau)d\tau \triangleq \bar{g} \tag{3.18}$$

and

$$\widehat{\Sigma}_g = \frac{1}{T}\sum_{t=1}^{T}(g(\tau_t)-\bar{g}_T)(g(\tau_t)-\bar{g}_T)' \longrightarrow \int_0^1 (g(\tau)-\bar{g})(g(\tau)-\bar{g})'d\tau \triangleq \Sigma_g, \tag{3.19}$$

where $\bar{g}$ is a finite vector with $||\bar{g}||_2 < \infty$ and Σ_g is a $k \times k$ positive definite matrix.

Assumption 3.3.1 regulates the properties of the weak trend components in the regressors. The positive definiteness of Σ_g rules out the possibility of multicollinearity among the k trend terms when $k > 1$. When $k = 1$, Σ_g becomes a positive value as long as $g(\tau)$ is not a constant on [0, 1]. This assumption ensures that the regression coefficients can be properly identified and estimated.

Assumption 3.3.2. (i) The error term v_t is strictly stationary with zero mean and finite variance. (ii) The error terms e_t and v_t are correlated with an unknown structure. Specifically, the conditional mean $\lambda(v) = \mathrm{E}(e_t|v_t = v)$ is a continuous and differentiable function defined on $\mathbb{R}^k \to \mathbb{R}^1$, whose functional form is unknown. (iii) The error term $u_t = e_t - \mathrm{E}(e_t|v_t)$ is a stationary α-mixing time series with mixing coefficient $\alpha(\cdot)$ satisfying $\sum_{d=1}^{\infty} \alpha^{\frac{\delta}{2+\delta}}(d) < \infty$.

Assumption 3.3.2 imposes standard requirements for the stationary α-mixing time series. We have assumed that the endogenous relationship between e_t and v_t is separable and can be represented by the control function of $\lambda(v_t)$.

The estimation process of our model involves two kernel functions. First is that when we estimate the conditional expectations with respect to v_t (or $\widehat{v}_t$ in practice), we adopt the k-dimensional kernel function $K_1(\cdot)$ with bandwidth h. The second is in the estimation of the nonparametric time trend $\widehat{g}(\tau)$, we employ a one-dimensional kernel function $K_2(\cdot)$ with bandwidth b.

Assumption 3.3.3. The kernel functions $K_1(\cdot)$ and $K_2(\cdot)$ are symmetric and continuous probability density functions satisfying $\int_{-\infty}^{\infty} K_i(u)du = 1$, $\int_{-\infty}^{\infty} uK_i(u)du = 0$, $\int_{-\infty}^{\infty} K_i(u)^2 du < \infty$, and $\int_{-\infty}^{\infty} uu'K_i(u)du$ is a positive definite matrix, for $i = 1, 2$.

Assumption 3.3.4. The two kernel functions $K_1(\cdot)$ and $K_2(\cdot)$ are associated with bandwidths h and b, respectively. As $T \to \infty$, they satisfy $h \to 0, b \to 0, Th^{2k} \to \infty$, $Th^2 \to \infty$, $Th^{5k} < \infty$, $Tb^5 < \infty$, and $b/h^k \to 0$.

The conditions in Assumption 3.3.3 can be easily satisfied, for example, the Epanechnikov kernel function $K(u) = 0.75(1 - u^2)$ $\mathbf{1}(|u| \leq 1)$, or the Gaussian kernel $K(u) = e^{-u^2/2}/\sqrt{2\pi}$. The conditions in Assumption 3.3.4 are also reasonable when the bandwidth is selected as the usual optimal bandwidth of order $O_p(T^{-0.2})$. Note that the bandwidth b has to converge to zero faster than h^k to ensure the consistency of the coefficient estimators.

A special feature in our estimation of the semiparametric partially linear model is that the smoothing variable v_t in the nonlinear component is unobservable. It is therefore replaced by its estimated value $\widehat{v}_t$ in the estimator. Hence, in all the following analysis, we follow a two-step procedure by first considering the properties of the infeasible estimator $\check{\beta}$ and then proving that the distance between the feasible estimator $\widehat{\beta}$ and the infeasible estimator $\check{\beta}$ is a small quantity that converges to zero in probability as $T \to \infty$. Thus, $\widehat{\beta}$ follows exactly the same asymptotic properties of $\breve{\beta}$ when the distance between them converges to zero sufficiently fast. The first theorem ensures that the estimator is well defined.

Lemma 3.3.1. *Under Assumptions* 3.3.1–3.3.4, *as* $T \to \infty$, *we have*

$$\breve{\Sigma}_T = \frac{1}{T}\sum_{t=1}^{T} \breve{x}_t \breve{x}_t' \longrightarrow_P \Sigma_g, \tag{3.20}$$

where Σ_g *is defined in Assumption* 3.3.1.

Remark 3.3.1. In fact, Σ_g can be regarded as a generalized version of the "variance–covariance matrix" that measures the variation in the time trends. Particularly, in the univariate case, a relatively flat time trend leads to a small value of Σ_g, which indicates relatively insufficient information in the trending component. Then it tends to cause relatively large standard errors in the regression coefficient estimators.

Theorem 3.3.1. *Under Assumptions* 3.3.1–3.3.4, *as* $T \to \infty$, *we have*

$$\sqrt{T}\left(\breve{\beta} - \beta\right) \longrightarrow_D \mathcal{N}(0, \Omega), \tag{3.21}$$

where $\Omega = \Sigma_g^{-1} \Lambda_u \Sigma_g^{-1'}$, *and* Λ_u *is the long-run variance of* $\breve{x}_t u_t$.

In the next step, we pass the asymptotic properties of the infeasible estimator to the feasible estimator by evaluating the distance between $\breve{\beta}$ and $\widehat{\beta}$. We first address in the following lemma that the distance between the information matrices is negligible.

Lemma 3.3.2. *Let* $\breve{\Sigma}_T = T^{-1}\sum_{t=1}^{T} \breve{x}_t \breve{x}_t'$ *and* $\widehat{\Sigma}_T = T^{-1}\sum_{t=1}^{T} \widehat{x}_t \widehat{x}_t'$. *Under Assumptions* 3.3.1–3.3.4, *as* $T \to \infty$, *we have*

$$\left|\left|\widehat{\Sigma}_T - \breve{\Sigma}_T\right|\right| = o_p(1). \tag{3.22}$$

Equation (3.22) is the key intermediate result we need to justify. According to the inequality

$$||\widehat{\Sigma}_T - \Sigma_g|| = ||\widehat{\Sigma}_T - \breve{\Sigma}_T + \breve{\Sigma}_T - \Sigma_g|| \leq ||\widehat{\Sigma}_T - \breve{\Sigma}_T|| + ||\breve{\Sigma}_T - \Sigma_g||, \tag{3.23}$$

Lemmas 3.3.1 and 3.3.2 show that both $||\widehat{\Sigma}_T - \check{\Sigma}_T||$ and $||\check{\Sigma}_T - \Sigma_g||$ converge to zero in probability. Therefore, we have the following:

Lemma 3.3.3. *Under Assumptions* 3.3.1–3.3.4, *as* $T \to \infty$, *we have*

$$\widehat{\Sigma}_T = \frac{1}{T}\sum_{t=1}^{T} \widehat{x}_t \widehat{x}_t' \longrightarrow_P \Sigma_g. \tag{3.24}$$

This lemma shows that the feasible information matrix also converges to Σ_g. We then proceed to the asymptotic properties of the feasible estimator. As in Theorem 3.3.1, we have shown that the infeasible estimator is asymptotically normal with convergence rate $\sqrt{T}$. The following lemma bounds the distance between the feasible and infeasible estimators.

Lemma 3.3.4. *Under Assumptions* 3.3.1–3.3.4, *as* $T \to \infty$, *we have*

$$\left|\left|\sqrt{T}(\widehat{\beta} - \check{\beta})\right|\right| = o_p(1). \tag{3.25}$$

Based on this lemma, the asymptotic normality for the feasible estimator $\widehat{\beta}$ is easily obtainable as follows.

Theorem 3.3.2. *Under Assumptions* 3.3.1–3.3.4, *as* $T \to \infty$, *we have*

$$\sqrt{T}\left(\widehat{\beta} - \beta\right) \longrightarrow_D \mathcal{N}(0, \Omega), \tag{3.26}$$

where $\Omega = \Sigma_g^{-1}\Lambda_u\Sigma_g^{-1'}$, *and* Λ_u *is the long-run variance of* $\check{x}_t u_t$.

Corollary 3.3.1. *Under Assumptions* 3.3.1–3.3.4, *as* $T \to \infty$, *we have*

$$\sqrt{T}\widehat{\Omega}^{-1/2}\left(\widehat{\beta} - \beta\right) \longrightarrow_D \mathcal{N}(0, I_k), \tag{3.27}$$

where I_k *is the* $k \times K$ *identity matrix and* $\widehat{\Omega} = \widehat{\Sigma}_T^{-1}\widehat{\Lambda}_u\widehat{\Sigma}_g^{-1'}$, *in which* $\widehat{\Sigma}_T = T^{-1}\sum_{t=1}^{T} \widehat{x}_t\widehat{x}_t'$ *and*

$$\widehat{\Lambda}_u = \sum_{l=-p}^{p} \omega_l \widehat{\Gamma}_L(l), \tag{3.28}$$

where $\widehat{\Gamma}_L(l)$ *is the* l^{th} *sample autocovariance of* $\widehat{x}_t\widehat{u}_t$ *and* $\widehat{u}_t = y_t - \widehat{\alpha} - x_t'\widehat{\beta}$. *For* $l = 1, 2, \ldots, p$, ω_l *is a weight function that guarantees*

$\widehat{\Gamma}_L(l)$ *nonnegative and* p *is a truncation parameter. For example, in the work of Phillips and Perron (1988), they used* $w_l = 1 - l/(p+1)$, *which was first proposed in the work of Newey and West (1987). Since both* $\widehat{\Sigma}_T$ *and* $\widehat{\Lambda}_u$ *are consistent estimators for* Σ_g *and* Λ_u, $\widehat{\Omega}$ *is a consistent estimator for* Ω.

Remark 3.3.2. Theorem 3.3.2 shows that the estimator $\widehat{\beta}$ is $\sqrt{T}$-consistent. The convergence rate is slower than the rate in the cointegration regression with $I(1)$ process due to the weak trend assumption. Since the weak trend $g(\tau_t)$ does not dominate the time series, the simple OLS estimator is therefore inconsistent due to endogeneity. This phenomenon is examined in the subsequent chapter of Monte Carlo simulations. Contrary to the limit distribution of the coefficients in the cointegration regressions, the limit distribution of $\sqrt{T}(\widehat{\beta} - \beta)$ is Gaussian with zero mean. Therefore, it is more convenient to conduct hypothesis tests for the parameters in (3.1) than for those in the cointegration models.

3.4 Implementation Steps and Computational Issues

In practice, we could estimate the trending regression model with endogeneity by the control function approach with the given dataset $\{y_t, x_{1t}, x_{2t}, \ldots, x_{kt}\}$ for $t = 1, 2, \ldots, T$. Suppose that the necessary assumptions in the previous sections are satisfied. To obtain estimates of the coefficients in (3.7), we follow the following estimation steps:

Step 1: We employ nonparametric kernel estimation methods[1] to estimate the weak trends $g_t(\tau_t) = (g_{1t}, g_{2t}, \ldots, g_{kt})'$ in $x_{it} = g_i(\tau_t) + v_{it}$ for $\tau_t = t/n$ and $i = 1, 2, \ldots, k$, $t = 1, 2, \ldots, T$. We denote the bandwidth used in the estimation as b. The nonparametric estimators for the trend functions are defined as

$$\widehat{g}_i(\tau_t) = \widehat{\mathrm{E}}_b[x_{it}|\tau_t], \tag{3.29}$$

[1]In R, a convenient way to apply nonparametric regression is to use the *np* package developed by Jeffery Racine and Tristen Hayfield. For details, see the manual at https://cran.r-project.org/web/packages/np/np.pdf.

for $t = 1, 2, \ldots, T$ and $i = 1, 2, \ldots, k$. Then, we compute the k residual sequences $\widehat{v}_{it} = x_{it} - \widehat{g}_i(\tau_t)$ as proxies for the real values of disturbances v_{it}, for $t = 1, 2, \ldots, T$ and $i = 1, 2, \ldots, k$.

Step 2: We compute the expectations of $y_t, x_{1t}, \ldots, x_{kt}$ conditional on the estimated values of the residual sequence $\widehat{v}_{1t}, \ldots, \widehat{v}_{kt}$ and denote them as $\widehat{y}_t(v)$ and $\widehat{x}_{it}(v)$. Specifically,

$$\widehat{y}_t(v) = \widehat{\mathrm{E}}_h[y_t|\widehat{v}_{1t}, \ldots, \widehat{v}_{kt}], \tag{3.30}$$

$$\widehat{x}_{it}(v) = \widehat{\mathrm{E}}_h[x_{it}|\widehat{v}_{1t}, \ldots, \widehat{v}_{kt}], \tag{3.31}$$

for $i = 1, 2, \ldots, k$ and $t = 1, 2, \ldots, T$. Note that for multivariate nonparametric regression, the kernel function is defined as the product of the kernel functions for each element, i.e.,

$$K(v_{1s}, \ldots, v_{ks}) = \prod_{i=1}^{k} K\left(\frac{v_{is} - v_{it}}{h}\right). \tag{3.32}$$

Then, the modified versions of the time series in equation (3.13) could be estimated by

$$\widehat{y}_t = y_t - \widehat{y}_t(v), \quad \widehat{x}_{it} = x_{it} - \widehat{x}_{it}(v), \tag{3.33}$$

for $i = 1, 2, \ldots, k$ and $t = 1, 2, \ldots, T$.

Step 3: Based on equation (3.15), we compute the estimated values of the coefficients:

$$\widehat{\beta} = (\widehat{\beta}_1, \ldots, \widehat{\beta}_k)' = \left(\sum_{t=1}^{T} \widehat{x}_t \widehat{x}_t'\right)^{-1} \sum_{t=1}^{T} \widehat{x}_t \widehat{y}_t. \tag{3.34}$$

Once the estimated values of $\beta_1, \ldots, \beta_k$ have been obtained, we could estimate the intercept term by

$$\widehat{\alpha} = \frac{1}{T} \sum_{t=1}^{T} (y_t - x_{1t}\widehat{\beta}_1 - \cdots - x_{kt}\widehat{\beta}_k), \tag{3.35}$$

and consequently, we could compute the regression residuals by $\widehat{e}_t = y_t - \widehat{\alpha} - x_{1t}\widehat{\beta}_1 - \cdots - x_{kt}\widehat{\beta}_k$.

Step 4: Since the endogenous correlation is captured by the control function $e_t = \lambda(v_{1t}, \ldots, v_{kt}) + u_t$, we can then recover it using the nonparametric kernel estimation method by

$$\widehat{\lambda}(v_t) = \widehat{\mathrm{E}}_h[\widehat{e}_t|\widehat{v}_{1t}, \ldots, \widehat{v}_{kt}]. \tag{3.36}$$

Remark 3.4.1. The estimation process involves nonparametric kernel estimation that we need to carefully select bandwidths of h and b. The selection of the bandwidths is the trade-off between the bias and the variance of the nonparametric estimates. In the literature, various bandwidth selection methods have been developed, such as the *rule-of-thumb and plug-in* method, the *cross-validation (CV)* method, and the *AIC type* methods, see the works of Härdle and Vieu (1992), Fan and Gijbels (1996), Fan and Yao (2003), Li and Racine (2007), and Cai (2007). Here, we apply the cross-validation method based on a grid search procedure. The optimal bandwidth of h for $\widehat{\mathrm{E}}_h[x_t|v_t]$ is the one that minimizes the objective function

$$h_{\text{opt}} = \arg\min_h \sum_{t=1}^{T} (x_t - \widehat{x}_{-1}(v_t, h))^2, \tag{3.37}$$

where $\widehat{x}_{-1}(v_t, h)$ is the leave-one-out kernel estimator of $\widehat{\mathrm{E}}_h[x_t|v_t]$. While for the optimal bandwidth b for the time trend estimation in $\widehat{\mathrm{E}}_b[x_t|\tau_t]$, as the error terms in (1.5) are allowed to be weakly dependent, we should apply a modified version of the cross-validation method by removing $2r + 1$ data points around x_t, i.e., we remove the data points from x_{t-r} to x_{t+r} to ensure that the autocorrelation in v_t does not affect the selection of b. Therefore, the optimal bandwidth b_{opt} is selected by

$$b_{\text{opt}} = \arg\min_b \sum_{t=1}^{T} (x_t - \widehat{g}_{-r}(\tau_t, b))^2, \tag{3.38}$$

and $\widehat{g}_{-r}(\tau_t, b)$ is the leave-$(2r + 1)$-out estimator based on the formula of $\widehat{\mathrm{E}}_b[x_t|\tau_t]$. Note that when the error terms are i.i.d., $r = 0$ is sufficient to eliminate the information at time t for cross-validation, i.e., the usual leave-one-out cross-validation method.

3.5 Simulated Example

In the above section, we proposed the estimator for the semi-parametric partially linear model, which is extended from the linear regression model by the nonparametric control function approach. In this section, we examine the performance of the proposed estimator in equation (3.15) with a comparison to the simple OLS estimator by using the simulated data. In every replication, we generate the time series data from the equations as follows:

$$y_t = \alpha + \beta_1 x_{1t} + \beta_2 x_{2t} + e_t, \tag{3.39}$$

$$x_{1t} = g_1(\tau_t) + v_{1t}, \tag{3.40}$$

$$x_{2t} = g_2(\tau_t) + v_{2t}, \tag{3.41}$$

where $\tau_t = t/T$ for $t = 1, 2, \ldots, T$. In the above process, we let $\alpha = 0.3, \beta_1 = 0.7, \beta_2 = 0.5$. The time trends are bounded weak trends of the form $g_1(\tau_t) = 3 - 4(\tau_t - 0.5)^2$ and $g_2(\tau_t) = 2 + 0.7\sin(2\pi\tau_t)$. The error term e_t is correlated with v_{1t} and v_{2t} that $e_t = 1.5v_{1t} + v_{2t} + u_t$. Meanwhile, v_{1t}, v_{2t}, and u_t follow stationary AR(1) processes $v_{it} = 0.2v_{i,t-1} + \eta_{it}$ and $u_t = 0.2u_{t-1} + \varepsilon_t$, where $(\varepsilon_t, \eta_{1t}, \eta_{2t})' \overset{i.i.d.}{\sim} \mathcal{N}(0, \Omega_3)$ in which $\Omega_3 = diag(0.2, 0.2, 0.2)$.

To show the effect of the control function approach, we present the behavior of the simple OLS estimator for β_1 and β_2 in (3.1) and the proposed estimator (3.15) in the semiparametric partially linear model

$$y_t = \alpha + x_{1t}\beta_1 + x_{2t}\beta_2 + \lambda(v_{1t}, v_{2t}) + u_t, \tag{3.42}$$

where $\lambda(\cdot)$ is treated as an unknown nonparametric control function and u_t is the error term independent with $x_{1t}, x_{2t}, v_{1t}, v_{2t}$.

The time series data are simulated independently for $N_B = 5000$ times. Each time, we conduct the simulation procedures and obtain the estimated values of $\widehat{\beta}_{i,p}^{\text{ols}}$ (the OLS estimator) and $\widehat{\beta}_{i,p}^{\text{control}}$ (the estimator in equation (3.15)), for $i = 1, 2$ and $p = 1, 2, \ldots, N_B$. Finally, we repeat the N_B times replication three times in which the sample sizes are chosen as 250, 500, and 1000, respectively.

To show the properties of the two estimators, we calculate the averages of the biases, the standard deviations, as well as the root mean squared errors for the two estimators based on the

Table 3.1. Simulation results for the weak trending regression with endogeneity.

		OLS			Control function		
	T	250	500	1000	250	500	1000
	Bias	0.4776	0.4755	0.4787	0.0105	0.0088	0.0028
$\widehat{\beta}_1$	Std	0.0663	0.0430	0.0340	0.1132	0.0764	0.0577
	RMSE	0.4822	0.4775	0.4799	0.1137	0.0769	0.0578
	Bias	0.1420	0.1446	0.1462	–0.0324	–0.0307	–0.0298
$\widehat{\beta}_2$	Std	0.0506	0.0307	0.0242	0.0719	0.0428	0.0343
	RMSE	0.1507	0.1479	0.1482	0.0789	0.0527	0.0454

N_B replications. The results are reported in Table 3.1. Specifically, they are computed by the following formulas:

$$\text{Bias}_i = \frac{1}{N_B}\sum_{p=1}^{N_B}\widehat{\beta}_{i,p} - \beta, \tag{3.43}$$

$$\text{Std}_i = \sqrt{\frac{1}{N_B-1}\sum_{p=1}^{N_B}\left(\widehat{\beta}_{i,p} - \frac{1}{N_B}\sum_{p=1}^{N_B}\widehat{\beta}_{i,p}\right)^2}, \tag{3.44}$$

$$\text{RMSE}_i = \sqrt{\text{Bias}_i^2 + \text{Std}_i^2}, \tag{3.45}$$

for $i = 1, 2$, and $\widehat{\beta}_{i,p}$ is replaced by $\widehat{\beta}_{i,p}^{\text{ols}}$ and $\widehat{\beta}_{i,p}^{\text{control}}$ for the two estimators.

According to the discussion in the previous chapters, since the weak trends do not dominate the stationary error terms, endogeneity causes persistent biases in the simple OLS estimators for the coefficients regardless of the sample size. Therefore, as expected, a nondiminishing positive bias is seen in the simple OLS estimates for β_1 (≈ 0.47) and $\beta_2(\approx 0.14)$. This result reconciles with the theoretical conclusion that the OLS estimators are inconsistent in the weakly trending regressions with endogeneity.

On the other hand, the control function extends the linear regression model to a semiparametric partially linear model as (3.42) so that we are able to solve the problem of endogeneity. We have shown that the proposed estimators for β_1 and β_2 converge to their true values when the sample size tends to infinity. As shown in Table 3.1, for the control function estimators, the biases are negligible, and the standard deviations decrease at the rate of $1/\sqrt{T}$ as $T \to \infty$, though they are slightly higher than those for OLS estimators. Therefore, the control function approach successfully adjusts for the endogeneity bias and yields unbiased and consistent estimates of the coefficients.

3.6 Empirical Example

In this section, we explore the long-run relationship between income and consumption expenditures. The data are the quarterly U.S. aggregate personal disposable income, aggregate personal consumption expenditure, and real interest rate from 1959Q1 to 2009Q4. The data can be downloaded from https://fred.stlouisfed.org/. The real interest rate is calculated as the difference between the nominal interest rate and the inflation rate.

The logarithms of aggregate personal disposable income and aggregate personal consumption expenditure are plotted in

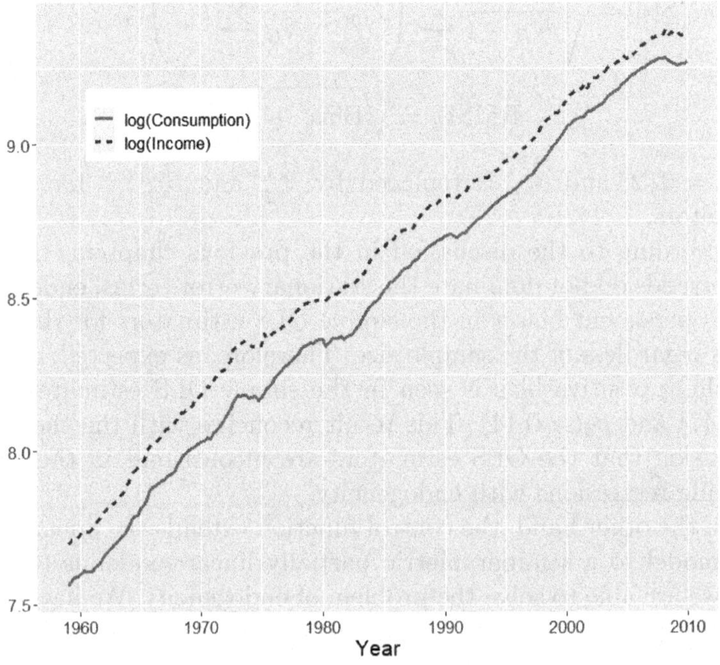

Figure 3.2. Logarithms of aggregate personal income and consumption.

Figure 3.2. We can find that the two time series track each other quite well. Both of them contain linear upward time trends, which show their long-term growth over the whole sample period. According to our definition, however, linear trends are strong trends and they are common trends by nature. To ensure that such a common linear trend does not affect the estimation of the long-run MPC, we estimate and then remove the linear time trends in both time series and denote their remainders as i_t and c_t for income and consumption, respectively.

Figure 3.3 shows the graphs of i_t (solid curve) and c_t (dashed curve) as well as the real interest rate r_t (dotted curve). These three time series are usually believed to be pure random walks in the literature because we usually fail to reject the null hypothesis of unit root using conventional unit-root tests. Based on the arguments in the previous chapters, however, the exact form of trend is usually unknown, and it is also reasonable to assume that these three sequences c_t, i_t, and r_t are nonlinear trend-stationary time series that the regressors

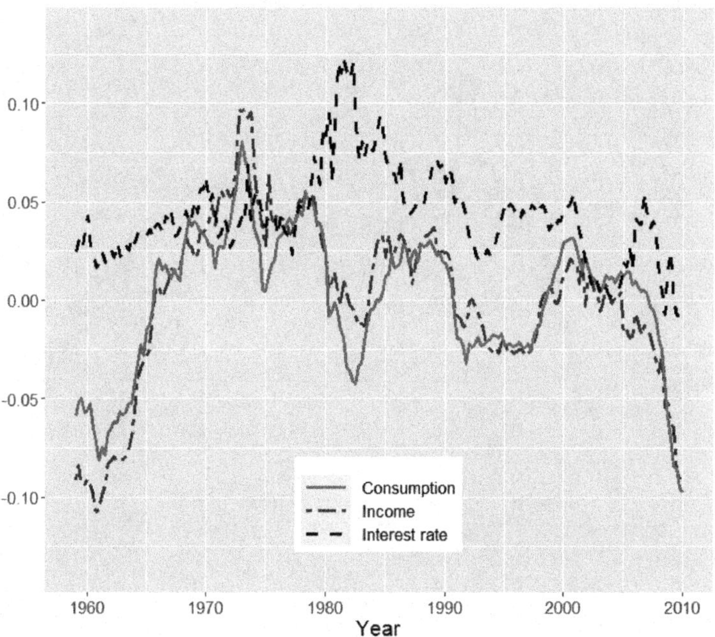

Figure 3.3. Consumption, income, and interest rate.

i_t and c_t could also be modeled as

$$i_t = g_1(\tau_t) + v_{1t}, \tag{3.46}$$

$$r_t = g_2(\tau_t) + v_{2t}, \tag{3.47}$$

where v_{it} is a sequence of stationary disturbances and $g_i(\tau_t)$ is a nonlinear and bounded function of deterministic time trend for $i = 1, 2$.

The two unknown functions $g_1(\tau_t)$ and $g_2(\tau_t)$ are estimated using nonparametric local linear kernel estimation methods as in Figures 3.4 and 3.5 with bandwidth $h = 0.08$. In Figures 3.4 and 3.5, the dashed curves are the estimated time trends, and the gray shaded areas indicate their estimated 95% confidence bands. Since a horizontal zeros line at any level could not be entirely contained in the 95% confidence bands, it is reasonable to conclude that nonlinear time trends exist in both i_t and r_t under 5% significance level.

Figure 3.6 shows the time series plot of the residual sequences $\widehat{v}_{1t} = i_t - \widehat{g}_1(\tau_t)$ and $\widehat{v}_{2t} = r_t - \widehat{g}_2(\tau_t)$, where $\widehat{g}_1(\tau_t)$ and $\widehat{g}_2(\tau_t)$ are the estimated nonlinear time trends in i_t and r_t. Table 3.2 reports the

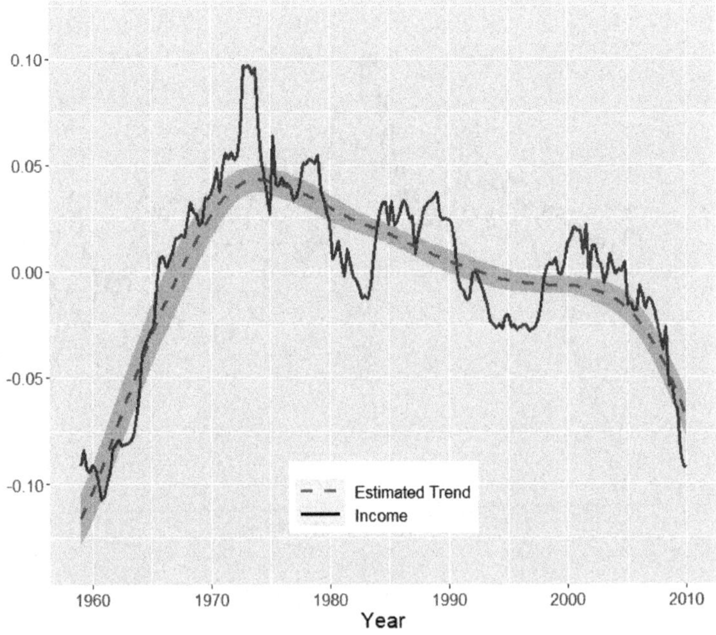

Figure 3.4. Income i_t and its estimated trend.

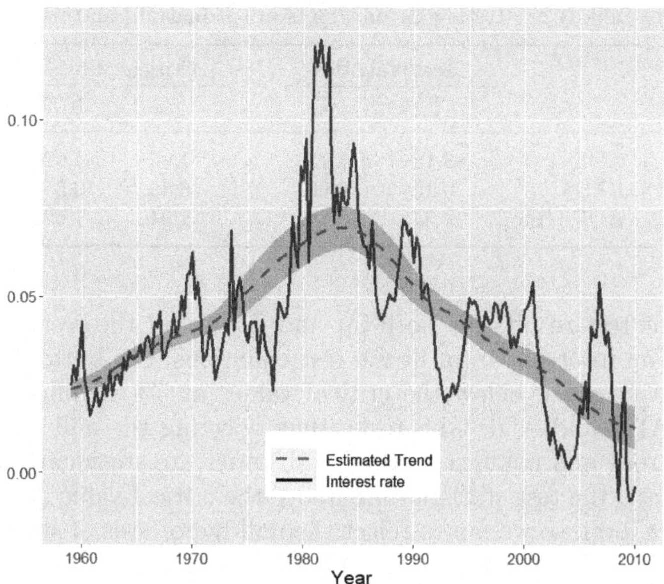

Figure 3.5. Interest rate r_t and its estimated trend.

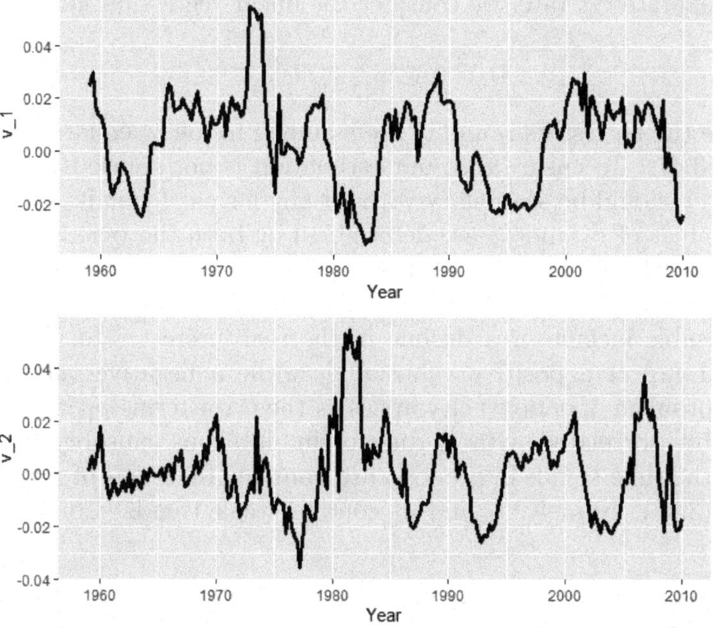

Figure 3.6. Time series plots of $\widehat{v}_{1t}$ and v_{2t}.

Table 3.2. Testing for unit roots in residuals $\widehat{v}_{1t}$ and $\widehat{v}_{2t}$.

	Test statistics		Critical values	
	$\widehat{v}_{1t}$	$\widehat{v}_{2t}$	5%	10%
ADF	–3.18	–3.73	–1.95	–1.62
KPSS	0.15	0.06	0.46	0.35
ADF-GLS	–2.34	–3.80	–1.94	–1.62

unit-root test results for both $\widehat{v}_{1t}$ and $\widehat{v}_{2t}$. In all the tests, we only include an intercept term in the test equations. For both series, the test statistics are below the critical values at 5% significance level of the ADF and ADF-GLS tests, thus rejecting the null hypothesis of unit root and confirming that both series are stationary. For the KPSS test, the test statistics are above the critical value at 10% significance level, so we fail to reject the null hypothesis of stationarity. Therefore, both $\widehat{v}_{1t}$ and $\widehat{v}_{2t}$ are stationary. Models (3.46) and (3.47) are one of the reasonable forms for i_t and r_t.

To reveal the long-run relationship between the aggregate personal disposable income, the personal consumption expenditure, and the real interest rate, we consider the linear regression model[2]

$$c_t = \alpha + i_t\beta_1 + r_t\beta_2 + e_t, \tag{3.48}$$

where the regressors i_t and r_t are assumed to follow equations (3.46) and (3.47). To ensure that our estimation is not spurious, the error term e_t should be stationary without any forms of trends. Therefore, the values of β_1 and β_2 are determined by both the weakly trending and the stationary components in i_t and r_t. In normal situations, consumers would spend more if there were an increase in personal disposable income or a decline in the real interest rate. Hence, we would expect a positive value of β_1 while a negative value of β_2. The problem of endogeneity arises as the error terms v_{1t} and v_{2t} are possibly correlated with e_t due to simultaneous causality. Suppose that the time trends in i_t and r_t are bounded functions of τ_t in (3.46) and (3.47). Namely, i_t and r_t contain weak trends. Consequently,

[2]It is the same to estimate $c_t = \alpha + \delta t + i_t\beta_1 + r_t\beta_2 + e_t$ if we do not remove linear trends in c_t and i_t.

the simple OLS estimator of the coefficients is likely to be biased and inconsistent, which may lead to unreliable economic conclusions.

To deal with the problem of endogeneity, we employ the nonparametric control function approach and define a two-dimensional nonparametric function as $e_t = \lambda(v_{1t}, v_{2t}) + u_t$, for $t = 1, 2, \ldots, T$, where $\lambda(v_{1t}, v_{2t}) = \mathrm{E}(e_t|v_{1t}, v_{2t})$ is a continuous and differentiable function defined on $\mathbb{R}^2 \to \mathbb{R}$ and $u_t = e_t - \mathrm{E}(e_t|v_{1t}, v_{2t})$ is the new error term that satisfies $\mathrm{E}(u_t|v_{1t}, v_{2t}) = 0$. Replacing e_t in the linear regression model, we have

$$c_t = \alpha + i_t\beta_1 + r_t\beta_2 + \lambda(v_{1t}, v_{2t}) + u_t, \tag{3.49}$$

in which the nonparametric control function captures the endogenous correlation without the risk of model misspecification. The problem of endogeneity disappears in the extended model since the new error term u_t is uncorrelated with i_t, r_t, v_{1t}, and v_{2t}. Note that as $\mathrm{E}(e_t) = 0$, we have $\mathrm{E}(\lambda(v_{1t}, v_{2t})) = 0$. To compare the performance of using a nonparametric control function and parametric alternatives, we also estimate equation (3.49), where the control function is specified as a linear function

$$\lambda(v_{1t}, v_{2t}) = \gamma_1 v_{1t} + \gamma_2 v_{2t} \tag{3.50}$$

and a nonlinear function

$$\lambda(v_{1t}, v_{2t}) = \gamma_1 v_{1t} + \gamma_2 v_{2t} + \gamma_{11} v_{1t}^2 + \gamma_{22} v_{2t}^2 + \gamma_{12} v_{1t} v_{2t}. \tag{3.51}$$

The estimation results of the coefficients for different models are summarized in Table 3.3. To illustrate the sensitivity of bandwidth selection, we report estimation results under different choices of bandwidth. Note that (b_1, b_2) are the bandwidths adopted to estimate $g_1(\tau_t)$ and $g_2(\tau_t)$, and (h_1, h_2) are the bandwidths to estimate the conditional means of $\mathrm{E}(y_t|v_{1t}, v_{2t})$ and $\mathrm{E}(x_t|v_{1t}, v_{2t})$. For the parametric control function approach, we also report the estimated values of γ_1 and γ_2 as in equations (3.50) and (3.51). The table shows that both $\widehat{\beta}_1$ and $\widehat{\beta}_2$ are significant at the 1% significance level. They also have the expected signs that higher aggregate personal income leads to higher consumption expenditures (therefore a positive β_1), while higher real interest rate encourages people to save more and spend less (therefore a negative β_2).

Table 3.3. Model estimation results.

		OLS				
	$\widehat{\beta}_1$	0.7976 (0.0223)				
	$\widehat{\beta}_2$	−0.2071 (0.0399)				
		Control function approach				
(b_1, b_2)		Parametric		Nonparametric (h_1, h_2)		
		Linear	Nonlinear	$(0.012, 0.012)$	$(0.024, 0.023)$	$(0.037, 0.035)$
$(0.06, 0.06)$	$\widehat{\beta}_1$	0.7359 (0.0311)	0.7431 (0.0303)	0.7380 (0.0304)	0.7435 (0.0301)	0.7390 (0.0305)
	$\widehat{\beta}_2$	−0.1986 (0.0671)	−0.1583 (0.0672)	−0.1498 (0.0682)	−0.1725 (0.0662)	−0.1834 (0.0663)
	$\widehat{\gamma}_1$	0.3247 (0.0760)	0.3547 (0.0769)			
	$\widehat{\gamma}_2$	0.1167 (0.0960)	0.1274 (0.0976)			
		Linear	Nonlinear	$(0.014, 0.012)$	$(0.028, 0.024)$	$(0.042, 0.036)$
$(0.08, 0.08)$	$\widehat{\beta}_1$	0.7217 (0.034)	0.7363 (0.0332)	0.7265 (0.0328)	0.733 (0.0328)	0.7278 (0.0335)
	$\widehat{\beta}_2$	−0.1952 (0.0758)	−0.1746 (0.0741)	−0.1651 (0.0739)	−0.1765 (0.0732)	−0.1859 (0.0744)
	$\widehat{\gamma}_1$	0.3196 (0.0679)	0.3392 (0.0690)			
	$\widehat{\gamma}_2$	0.1376 (0.0978)	0.1816 (0.0983)			
		Linear	Nonlinear	$(0.016, 0.013)$	$(0.033, 0.026)$	$(0.048, 0.039)$
$(0.10, 0.10)$	$\widehat{\beta}_1$	0.7139 (0.0374)	0.7427 (0.0362)	0.7324 (0.0360)	0.7327 (0.0362)	0.7259 (0.0368)
	$\widehat{\beta}_2$	−0.2024 (0.0823)	−0.2016 (0.0787)	−0.1956 (0.0786)	−0.1906 (0.0785)	−0.1993 (0.0802)
	$\widehat{\gamma}_1$	0.2939 (0.0632)	0.3084 (0.0639)			
	$\widehat{\gamma}_2$	0.1537 (0.1008)	0.2288 (0.1010)			

The simple OLS estimate for β_1 is 0.798 with a standard error of 0.022. This OLS estimate is higher than those with control functions that adjust for the problem of endogeneity with either parametric or nonparametric forms. After taking into account the issue of endogeneity, the estimated MPC is around 0.73 with a standard error around 0.03. This estimate is robust to bandwidth choices in this example because, under different bandwidths, the estimated values of β_1 vary within their standard errors. The OLS method overestimates the elasticity of income to consumption as the two variables affect each other simultaneously. This is in accordance with the fact that the estimated value of γ_1 is significantly positive when using parametric control functions. For both OLS and control function approaches, the estimated values of β_2 are not significantly different indicating that the real interest rate may be exogenous. This coincides with the insignificance of γ_2.

Based on $\widehat{\beta}_1$ and $\widehat{\beta}_2$ using the nonparametric control function approach, we are able to recover the unknown control function $\lambda(v_{1t}, v_{2t})$ based on the estimated residuals of $\widehat{e}_t, \widehat{v}_{1t}, \widehat{v}_{2t}$. Figure 3.7 is

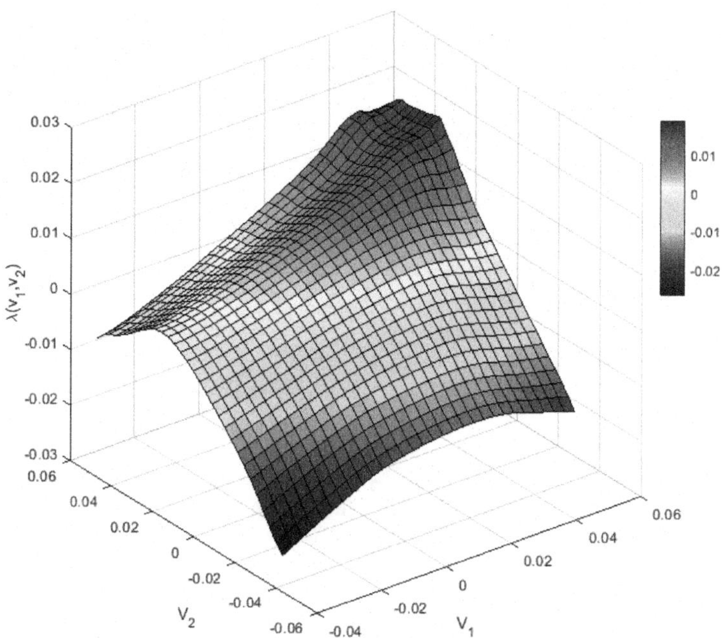

Figure 3.7. The local linear kernel estimation of the nonparametric control function $\lambda(v_{1t}, v_{2t})$.

a 3D plot of the local linear kernel estimates of the control function. The estimated function shows nonlinear characteristics that can hardly be expressed as simple functions of v_{1t} and v_{2t}. In other words, an incorrect parametric specification for the control function, for example, a simple flat surface function $\lambda(v_{1t}, v_{2t}) = \gamma_1 v_{1t} + \gamma_2 v_{2t}$, could not fully capture the endogenous relationship between e_t and (v_{1t}, v_{2t}). As a result, we are still likely to obtain biased and inconsistent estimates of the MPC.

In addition, Figures 3.8 and 3.9 present the nonparametric kernel estimates of the expectations of the control function conditional on v_{1t} and v_{2t}, respectively. The 2D graphs are simply projections of the surface in Figure 3.7 and provide much convenience to the determination of the significance of the control functions with respect to v_{1t} and v_{2t}. The 95% confidence bands (the gray area) show that both regressors are endogenously correlated with the error term in

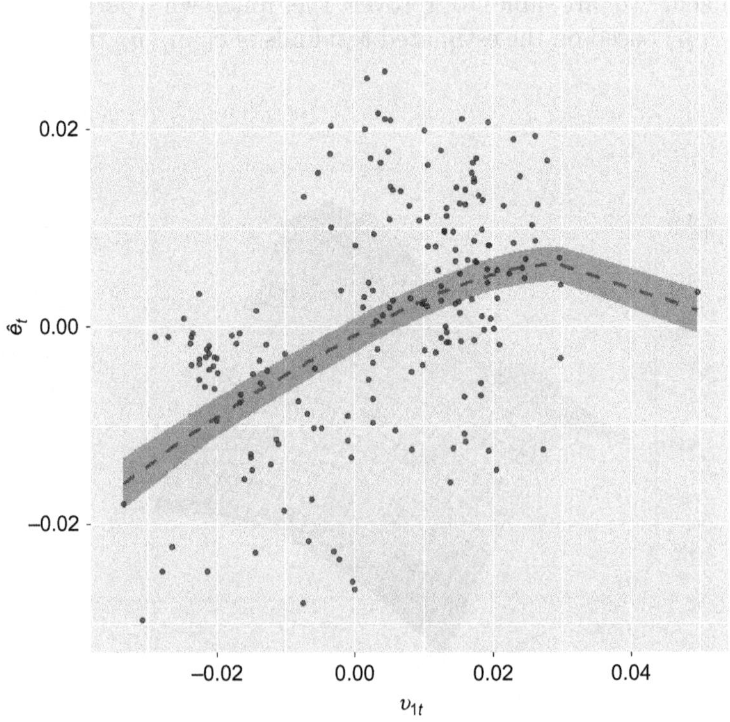

Figure 3.8. $m_1(v_1) = \mathrm{E}[\lambda(v_{1t}, v_{2t})|v_{1t} = v_1]$.

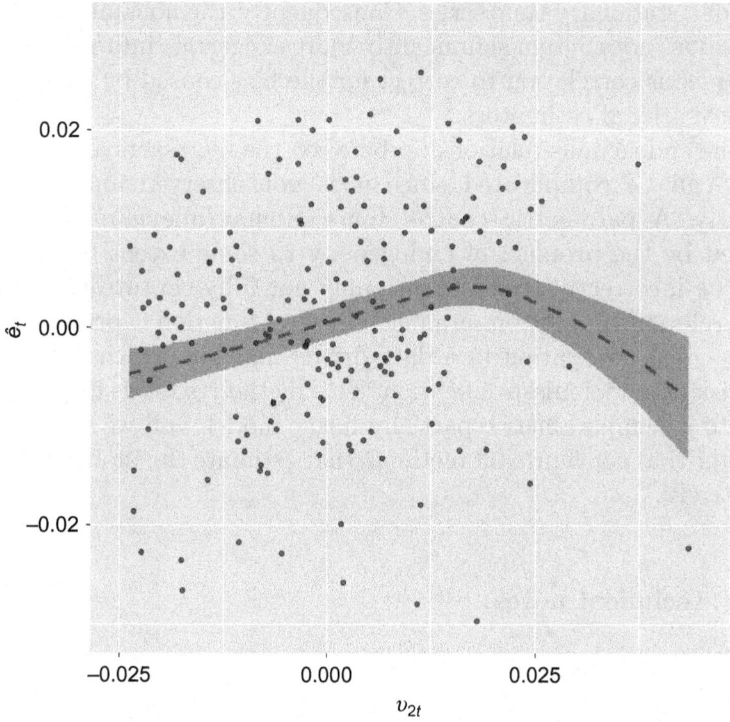

Figure 3.9. $m_2(v_2) = \mathrm{E}[\lambda(v_{1t}, v_{2t})|v_{2t} = v_2]$.

the regression model (3.1). Also, the form of such correlation is not linear as what we have observed in the graph of $\widehat{\lambda}(\widehat{v}_{1t}, \widehat{v}_{2t})$ in Figure 3.7. The two figures also show that the correlation between $\widehat{e}_t$ and $\widehat{v}_{1t}$ is stronger than that between $\widehat{e}_t$ and $\widehat{v}_{2t}$, which is very close to zero.[3]

3.7 Summary

In a nonstationary time series with a weak trend, the trend does not completely dominate the entire time series. Specifically, the weak time trend is limited and does not increase indefinitely. Additionally, the square summation of the weak trend is similar in magnitude to

[3]In fact, the OLS linear regression of $\widehat{e}_t$ over $\widehat{v}_{2t}$ is insignificant.

that of a stationary time series. Consequently, the nonstationary time series does not contain significantly more exogenous information than endogenous correlation to correct for the bias caused by endogeneity in conventional estimators.

The endogenous relationship between the regressors and the error term can be complicated since it is not observed and measured directly. A parametric control function may alleviate the impact caused by the problem of endogeneity to some extent. However, it may be incorrectly specified and may not fully capture the endogenous relationship. A nonparametric control function captures the real endogenous correlation in a data-driven manner, which is free from the risk of model misspecification. This method extends linear regression to a semiparametric partially linear model, and we show in this chapter that conventional methods that estimate the model still work effectively.

3.8 Technical notes

Some useful Lemmas

We first introduce the following lemmas that would provide much convenience to the proofs of the theorems in this chapter.

Lemma 3.8.1. *Under Assumptions* 3.3.1–3.3.4, *as* $T \to \infty$, *let* $\bar{g} = \int_0^1 g(\tau)d\tau$, $\bar{g}_i = \int_0^1 g_i(\tau)d\tau$, $\breve{g}_i(\tau_t) = g_i(\tau_t) - \sum_{s=1}^T w_{Ts}(t)g_i(\tau_s)$, $\breve{x}_{it} = x_{it} - \sum_{s=1}^T w_{Ts}(t)x_{is}$. *We define*

$$M_1(i,j) = \frac{1}{T}\sum_{t=1}^T (g_i(\tau_t) - \bar{g}_{i,T})(g_j(\tau_t) - \bar{g}_{j,T}) \longrightarrow \int_0^1 (g_i(\tau) - \bar{g}_i)(g_j(\tau) - \bar{g}_j)d\tau, \tag{3.52}$$

$$M_2(i,j) = \frac{1}{T}\sum_{t=1}^T \left(\bar{g}_{i,T} - \sum_{s=1}^T w_{Ts}(t)g_i(\tau_s)\right)\left(\bar{g}_{j,T} - \sum_{s=1}^T w_{Ts}(t)g_j(\tau_s)\right) \longrightarrow_P 0, \tag{3.53}$$

$$M_{12}(i,j) = \frac{1}{T}\sum_{t=1}^{T}(g_i(\tau_t) - \bar{g}_{i,T})\left(\bar{g}_{j,T} - \sum_{s=1}^{T} w_{Ts}(t) g_j(\tau_s)\right) \longrightarrow_P 0, \tag{3.54}$$

$$M_{21}(i,j) = \frac{1}{T}\sum_{t=1}^{T}\left(\bar{g}_{i,T} - \sum_{s=1}^{T} w_{Ts}(t) g_i(\tau_s)\right)(g_j(\tau_t) - \bar{g}_{j,T}) \longrightarrow_P 0, \tag{3.55}$$

$$S_2(i,j) = \frac{1}{T}\sum_{t=1}^{T}\left(v_{it} - \sum_{s=1}^{T} w_{Ts}(t) v_{is}\right)\left(v_{jt} - \sum_{s=1}^{T} w_{Ts}(t) v_{js}\right) \longrightarrow_P 0. \tag{3.56}$$

Lemma 3.8.2. *Under Assumptions* 3.3.1–3.3.4, *as* $T \to \infty$, *let* $\zeta(v) = (\partial\lambda/\partial v_1, \ldots, \partial\lambda/\partial v_k)'$ *be the first-order derivative of* $\lambda(v)$ *with respect to vector* v, *then*

$$I_{1T} = \left\| \frac{1}{\sqrt{T}} \sum_{t=1}^{T} \breve{v}_t \zeta'(v_t) \breve{v}_t \right\| = o_p(1) \tag{3.57}$$

and

$$I_{2T} = \left\| \frac{1}{\sqrt{T}} \sum_{t=1}^{T} \breve{g}(\tau_t) \zeta'(v_t) \breve{v}_t \right\| = o_p(1). \tag{3.58}$$

Lemma 3.8.3. *Under Assumptions* 3.3.1–3.3.4, *as* $T \to \infty$,

$$\left\| \frac{1}{\sqrt{T}} \sum_{t=1}^{T} \breve{x}_t \bar{u}_t \right\| = o_p(1), \tag{3.59}$$

where $\bar{u}_t = \sum_{s=1}^{T} w_{Ts}(t) u_s$.

Lemma 3.8.4. *Under Assumptions* 3.3.1–3.3.4, *as* $T \to \infty$,

$$||\mathfrak{D}_{1T}|| = \left\| \frac{1}{T} \sum_{t=1}^{T} (\widehat{x}_t - \breve{x}_t)(\widehat{x}_t - \breve{x}_t)' \right\| = o_p(1). \tag{3.60}$$

Lemma 3.8.5. *Let Assumptions* 3.3.1–3.3.4 *hold. As* $T \to \infty$, *we have*

$$\left\| \frac{\widehat{X}'\widehat{e} - \check{X}'\check{e}}{\sqrt{T}} \right\| = o_p(1). \tag{3.61}$$

Proof of Lemma 3.3.1. Recall that x_t is a k-dimensional vector of trending time series sequence, hence $\check{\Sigma}_T$ is a $k \times k$ matrix. To prove $\check{\Sigma}_T \longrightarrow_P \Sigma_g$, it suffices to show that $\check{\Sigma}_T(i,j) \longrightarrow_P \Sigma_g(i,j)$, for $i,j = 1,\ldots,k$. Note that $\check{x}_{it} = x_{it} - \sum_{s=1}^T w_{Ts}(t)x_{is}$ and $x_{it} = g_i(\tau_t) + v_{it}$, therefore,

$$\begin{aligned}
\check{\Sigma}_T(i,j) &= \frac{1}{T}\sum_{t=1}^T \check{x}_{it}\check{x}_{jt} = \frac{1}{T}\sum_{t=1}^T \left(x_{it} - \sum_{s=1}^T w_{Ts}(t)x_{is}\right)\left(x_{jt} - \sum_{s=1}^T w_{Ts}(t)x_{js}\right)\\
&= \frac{1}{T}\sum_{t=1}^T \left(g_i(\tau_t) - \sum_{s=1}^T w_{Ts}(t)g_i(\tau_s) + v_{it} - \sum_{s=1}^T w_{Ts}(t)v_{is}\right)\\
&\quad \times \left(g_j(\tau_t) - \sum_{s=1}^T w_{Ts}(t)g_j(\tau_s) + v_{jt} - \sum_{s=1}^T w_{Ts}(t)v_{js}\right)\\
&= \frac{1}{T}\sum_{t=1}^T \left(g_i(\tau_t) - \sum_{s=1}^T w_{Ts}(t)g_i(\tau_s)\right)\left(g_j(\tau_t) - \sum_{s=1}^T w_{Ts}(t)g_j(\tau_s)\right)\\
&\quad + \frac{1}{T}\sum_{t=1}^T \left(v_{it} - \sum_{s=1}^T w_{Ts}(t)v_{is}\right)\left(v_{jt} - \sum_{s=1}^T w_{Ts}(t)v_{js}\right)\\
&\quad + \frac{1}{T}\sum_{t=1}^T \left(g_i(\tau_t) - \sum_{s=1}^T w_{Ts}(t)g_i(\tau_s)\right)\left(v_{jt} - \sum_{s=1}^T w_{Ts}(t)v_{js}\right)\\
&\quad + \frac{1}{T}\sum_{t=1}^T \left(v_{it} - \sum_{s=1}^T w_{Ts}(t)v_{is}\right)\left(g_j(\tau_t) - \sum_{s=1}^T w_{Ts}(t)g_j(\tau_s)\right)\\
&\triangleq \frac{1}{T}\sum_{t=1}^T \breve{g}_i(\tau_t)\breve{g}_j(\tau_t) + \frac{1}{T}\sum_{t=1}^T \check{v}_{it}\check{v}_{jt} + \frac{1}{T}\sum_{t=1}^T \breve{g}_i(\tau_t)\check{v}_{jt} + \frac{1}{T}\sum_{t=1}^T \check{v}_{it}\breve{g}_j(\tau_t)\\
&\triangleq S_1(i,j) + S_2(i,j) + S_{12}(i,j) + S_{21}(i,j).
\end{aligned} \tag{3.62}$$

In the above equations, we define

$$S_1(i,j) = \frac{1}{T}\sum_{t=1}^{T} \breve{g}_i(\tau_t)\breve{g}_j(\tau_t), \quad S_2(i,j) = \frac{1}{T}\sum_{t=1}^{T} \breve{v}_{it}\breve{v}_{jt},$$

$$S_{12}(i,j) = \frac{1}{T}\sum_{t=1}^{T} \breve{g}_i(\tau_t)\breve{v}_{jt}, \quad S_{21}(i,j) = \frac{1}{T}\sum_{t=1}^{T} \breve{v}_{it}\breve{g}_j(\tau_t), \tag{3.63}$$

where $\breve{v}_{it} = v_{it} - \sum_{s=1}^{T} w_{Ts}(t)v_{is}$, for $i,j = 1,2,\ldots,k$, $t = 1,2,\ldots,T$. Let $\bar{g}_{i,T} = T^{-1}\sum_{t=1}^{T} g_i(\tau_t)$ for $i = 1,2,\ldots,k$. Therefore, $\bar{g}_{i,T}$ denotes the sample average of the trend component of x_i. We further decompose $S_1(i,j)$ as

$$\begin{aligned}
S_1(i,j) &= \frac{1}{T}\sum_{t=1}^{T}\left(g_i(\tau_t) - \sum_{s=1}^{T} w_{Ts}(t)g_i(\tau_s)\right) \\
&\qquad \times \left(g_j(\tau_t) - \sum_{s=1}^{T} w_{Ts}(t)g_j(\tau_s)\right) \\
&= \frac{1}{T}\sum_{t=1}^{T}\left(g_i(\tau_t) - \bar{g}_{i,T} + \bar{g}_{i,T} - \sum_{s=1}^{T} w_{Ts}(t)g_i(\tau_s)\right) \\
&\qquad \left(g_j(\tau_t) - \bar{g}_{j,T} + \bar{g}_{j,T} - \sum_{s=1}^{T} w_{Ts}(t)g_j(\tau_s)\right) \\
&= \frac{1}{T}\sum_{t=1}^{T}\left(g_i(\tau_t) - \bar{g}_{i,T}\right)\left(g_j(\tau_t) - \bar{g}_{j,T}\right) \\
&\quad + \frac{1}{T}\sum_{t=1}^{T}\left(\bar{g}_{i,T} - \sum_{s=1}^{T} w_{Ts}(t)g_i(\tau_s)\right)\left(\bar{g}_{j,T} - \sum_{s=1}^{T} w_{Ts}(t)g_j(\tau_s)\right) \\
&\quad + \frac{1}{T}\sum_{t=1}^{T}\left(g_i(\tau_t) - \bar{g}_{i,T}\right)\left(\bar{g}_{j,T} - \sum_{s=1}^{T} w_{Ts}(t)g_j(\tau_s)\right) \\
&\quad + \frac{1}{T}\sum_{t=1}^{T}\left(\bar{g}_{i,T} - \sum_{s=1}^{T} w_{Ts}(t)g_i(\tau_s)\right)\left(g_j(\tau_t) - \bar{g}_{j,T}\right) \\
&\triangleq M_1(i,j) + M_2(i,j) + M_{12}(i,j) + M_{21}(i,j),
\end{aligned} \tag{3.64}$$

where

$$M_1(i,j) = \frac{1}{T}\sum_{t=1}^{T}\left(g_i(\tau_t) - \bar{g}_{i,T}\right)\left(g_j(\tau_t) - \bar{g}_{j,T}\right), \tag{3.65}$$

$$M_2(i,j) = \frac{1}{T}\sum_{t=1}^{T}\left(\bar{g}_{i,T} - \sum_{s=1}^{T} w_{Ts}(t) g_i(\tau_s)\right)\left(\bar{g}_{j,T} - \sum_{s=1}^{T} w_{Ts}(t) g_j(\tau_s)\right), \tag{3.66}$$

$$M_{12}(i,j) = \frac{1}{T}\sum_{t=1}^{T}\left(g_i(\tau_t) - \bar{g}_{i,T}\right)\left(\bar{g}_{j,T} - \sum_{s=1}^{T} w_{Ts}(t) g_j(\tau_s)\right), \tag{3.67}$$

$$M_{21}(i,j) = \frac{1}{T}\sum_{t=1}^{T}\left(\bar{g}_{i,T} - \sum_{s=1}^{T} w_{Ts}(t) g_i(\tau_s)\right)\left(g_j(\tau_t) - \bar{g}_{j,T}\right). \tag{3.68}$$

Therefore, by Lemma 3.8.1,

$$\begin{aligned} S_1(i,j) &= M_1(i,j) + M_2(i,j) + M_{12}(i,j) + M_{21}(i,j) \\ &\longrightarrow_P \int_0^1 (g_i(\tau) - \bar{g}_i)(g_j(\tau) - \bar{g}_j)d\tau, \end{aligned} \tag{3.69}$$

$$S_2(i,j) \longrightarrow_P 0. \tag{3.70}$$

Meanwhile, by Cauchy–Schwarz inequality, given (3.69) and (3.70),

$$\begin{aligned} |S_{12}(i,j)| &= \left|\frac{1}{T}\sum_{t=1}^{T}\left(g_i(\tau_t) - \sum_{s=1}^{T} w_{Ts}(t) g_i(\tau_s)\right)\left(v_{jt} - \sum_{s=1}^{T} w_{Ts}(t) v_{js}\right)\right| \\ &\leq \left|\frac{1}{T}\sum_{t=1}^{T}\left(g_i(\tau_t) - \sum_{s=1}^{T} w_{Ts}(t) g_i(\tau_s)\right)^2\right|^{1/2} \left|\frac{1}{T}\sum_{t=1}^{T}\left(v_{jt} - \sum_{s=1}^{T} w_{Ts}(t) v_{js}\right)^2\right|^{1/2} \\ &= |S_1(i,i)|^{1/2}|S_2(j,j)|^{1/2} = O_p(1)O_p(1) = O_p(1). \end{aligned} \tag{3.71}$$

The same result holds for $S_{21}(i,j)$. Hence, for $i,j = 1,\ldots,k$,

$$\frac{1}{T}\sum_{t=1}^{T}\check{x}_{it}\check{x}_{jt} \longrightarrow_P \int_0^1 (g_i(\tau) - \bar{g}_i)(g_j(\tau) - \bar{g}_j)d\tau, \tag{3.72}$$

i.e., $\check{\Sigma}_T(i,j) \longrightarrow_P \Sigma_g(i,j)$. Therefore, the convergence of every element in $\check{\Sigma}_T$ yields the convergence of the whole matrix that as $T \to \infty$, $\check{\Sigma}_T \longrightarrow_P \Sigma_g$. ■

Proof of Lemma 3.3.4. Note that

$$\sqrt{T}B_T = \left(\frac{1}{T}\sum_{t=1}^{T}\check{x}_t\check{x}_t'\right)^{-1}\left(\frac{1}{\sqrt{T}}\sum_{t=1}^{T}\check{x}_t\check{\lambda}(v_t)\right). \tag{3.73}$$

Theorem 3.3.1 shows $T^{-1}\sum_{t=1}^{T}\check{x}_t\check{x}_t' \longrightarrow_P \Sigma_g$, hence $\left(T^{-1}\sum_{t=1}^{T}\check{x}_t\check{x}_t'\right)^{-1}$ converges in probability to Σ_g^{-1}, which is $O_p(1)$. Without loss of generality, we assume that $k = 1$. Our objective is to show that

$$\frac{1}{\sqrt{T}}\sum_{t=1}^{T}\check{x}_t\check{\lambda}(v_t) = O_p(1). \tag{3.74}$$

Note that $\check{x}_t$ can be written as

$$\begin{aligned}\check{x}_t &= x_t - \sum_{s=1}^{T}w_{Ts}(t)x_s = v_t + g(\tau_t) - \sum_{s=1}^{T}w_{Ts}(t)v_s - \sum_{s=1}^{T}w_{Ts}(t)g(\tau_s)\\ &= v_t - \sum_{s=1}^{T}w_{Ts}(t)v_s + g(\tau_t) - \sum_{s=1}^{T}w_{Ts}(t)g(\tau_s) = \check{v}_t + \check{g}(\tau_t),\end{aligned} \tag{3.75}$$

where $\check{v}_t = v_t - \sum_{s=1}^{T}w_{Ts}(t)v_s$ and $\check{g}(\tau_t) = g(\tau_t) - \sum_{s=1}^{T}w_{Ts}(t)g(\tau_s)$. Meanwhile, $\check{\lambda}(v_t)$ can be written as

$$\check{\lambda}(v_t) = \lambda(v_t) - \sum_{s=1}^{T}w_{Ts}(t)\lambda(v_s) = \sum_{s=1}^{T}w_{Ts}(t)\left(\lambda(v_t) - \lambda(v_s)\right). \tag{3.76}$$

Apply the Taylor expansion for $\lambda(v_s)$,

$$\lambda(v_s) = \lambda(v_t) + \lambda^{(1)}(v_t)(v_s - v_t) + \frac{1}{2}\lambda^{(2)}(v_t^*)(v_s - v_t)^2, \tag{3.77}$$

where v_t^* lies between v_t and v_s, and $\lambda^{(2)}(v_t^*) < \infty$ for any t. Therefore,

$$
\begin{aligned}
\check{\lambda}(v_t) &= \sum_{s=1}^{T} w_{Ts}(t)\left(\lambda(v_t) - \lambda(v_s)\right) \\
&= \sum_{s=1}^{T} w_{Ts}(t)\left(\lambda^{(1)}(v_t)(v_t - v_s) - \frac{1}{2}\lambda^{(2)}(v_t^*)(v_s - v_t)^2\right) \\
&= \lambda^{(1)}(v_t)\sum_{s=1}^{T} w_{Ts}(t)(v_t - v_s) - \frac{1}{2}\lambda^{(2)}(v_t^*)\sum_{s=1}^{T} w_{Ts}(t)(v_t - v_s)^2 \\
&= \lambda^{(1)}(v_t)\check{v}_t - \frac{1}{2}\lambda^{(2)}(v_t^*)\sum_{s=1}^{T} w_{Ts}(t)(v_s - v_t)^2. \qquad (3.78)
\end{aligned}
$$

Substitute $\check{x}_t$ in (3.74) with (3.75) and $\check{\lambda}(v_t)$ with its leading term in (3.78) and denote the first-order derivative of $\lambda(v_t)$ as $\zeta(v_t)$:

$$
\begin{aligned}
\frac{1}{\sqrt{T}}\sum_{t=1}^{T}\check{x}_t\check{\lambda}(v_t) &= \frac{1}{\sqrt{T}}\sum_{t=1}^{T}\check{v}_t\check{\lambda}(v_t) + \frac{1}{\sqrt{T}}\sum_{t=1}^{T}\check{g}(\tau_t)\check{\lambda}(v_t) \\
&= \left(\frac{1}{\sqrt{T}}\sum_{t=1}^{T}\zeta(v_t)\check{v}_t^2 + \frac{1}{\sqrt{T}}\sum_{t=1}^{T}\check{g}(\tau_t)\zeta(v_t)\check{v}_t\right) \\
&\quad \times (1 + o_p(1)) \\
&= (I_{1T} + I_{2T})(1 + o_p(1)), \qquad (3.79)
\end{aligned}
$$

where we defined I_{1T} as $\frac{1}{\sqrt{T}}\sum_{t=1}^{T}\zeta(v_t)\check{v}_t^2$ and I_{2T} as $\frac{1}{\sqrt{T}}\sum_{t=1}^{T}\check{g}(\tau_t)\zeta(v_t)\check{v}_t$. Therefore, we complete the proof provided that Lemma 3.8.2 holds. ■

Proof of Theorem 3.3.1. Substituting $\check{y}_t$ in $\check{\beta}$ using $\check{y}_t = \check{x}_t'\beta + \check{\lambda}(v_t) + \check{u}_t$, we have

$$\begin{aligned}
\check{\beta} &= \left(\sum_{t=1}^{T} \check{x}_t\check{x}_t'\right)^{-1}\left(\sum_{t=1}^{T} \check{x}_t\left(\check{x}_t'\beta + \check{\lambda}(v_t) + \check{u}_t\right)\right) \\
&= \beta + \left(\sum_{t=1}^{T} \check{x}_t\check{x}_t'\right)^{-1}\left(\sum_{t=1}^{T} \check{x}_t\check{\lambda}(v_t)\right) + \left(\sum_{t=1}^{T} \check{x}_t\check{x}_t'\right)^{-1}\left(\sum_{t=1}^{T} \check{x}_t\check{u}_t\right) \\
&= \beta + B_T + \left(\sum_{t=1}^{T} \check{x}_t\check{x}_t'\right)^{-1}\left(\sum_{t=1}^{T} \check{x}_t\check{u}_t\right), \qquad (3.80)
\end{aligned}$$

where $B_T = \left(\sum_{t=1}^{T} \check{x}_t\check{x}_t'\right)^{-1}\left(\sum_{t=1}^{T} \check{x}_t\check{\lambda}(v_t)\right)$. Therefore,

$$\sqrt{n}\left(\check{\beta} - \beta - B_T\right) = \left(\frac{1}{T}\sum_{t=1}^{T} \check{x}_t\check{x}_t'\right)^{-1}\left(\frac{1}{\sqrt{T}}\sum_{t=1}^{T} \check{x}_t\check{u}_t\right). \qquad (3.81)$$

In the previous proofs of Theorem 3.3.1, we have shown that

$$\frac{1}{T}\sum_{t=1}^{T} \check{x}_t\check{x}_t' \longrightarrow_P \Sigma_g, \qquad (3.82)$$

where Σ_g is assumed to be positive definite, therefore invertible. For the latter part,

$$\frac{1}{\sqrt{T}}\sum_{t=1}^{T} \check{x}_t\check{u}_t = \frac{1}{\sqrt{T}}\sum_{t=1}^{T} \check{x}_t(u_t - \bar{u}_t) = \frac{1}{\sqrt{T}}\sum_{t=1}^{T} \check{x}_t u_t + \frac{1}{\sqrt{T}}\sum_{t=1}^{T} \check{x}_t\bar{u}_t, \qquad (3.83)$$

where $\bar{u}_t = \sum_{s=1}^{T} w_{Ts}(t)u_s$. Under Assumptions 3.3.1–3.3.4, since u_t is mixing and independent with v_t, by Central Limit Theorem (CLT) for mixing processes (see the work of Fan and Yao (2003), Theorem 2.21), as $T \to \infty$,

$$\frac{1}{\sqrt{T}} \sum_{t=1}^{T} \check{x}_t u_t \longrightarrow_D \mathcal{N}(0, \Sigma_g \Lambda_u), \tag{3.84}$$

where Λ_u is the long-run variances of u_t:

$$\Lambda_u = \mathrm{E}\left[u_t u_t\right] + 2\sum_{j=1}^{\infty} \mathrm{E}\left[u_t u_{t-j}\right]. \tag{3.85}$$

Meanwhile, by Lemma 3.8.3, we have

$$\left\| \frac{1}{\sqrt{T}} \sum_{t=1}^{T} \check{x}_t \bar{u}_t \right\| = o_p(1). \tag{3.86}$$

Then, by the Slutsky theorem, we have

$$\sqrt{n}\left(\check{\beta} - \beta - B_T\right) \longrightarrow_D \mathcal{N}(0, \Omega), \tag{3.87}$$

where $\Omega = \Sigma_g^{-1}\Lambda_u$. By Lemma 3.3.4, the bias term B_T is negligible that $\sqrt{T}B_T = o_p(1)$ as $T \to \infty$. Thus, we can ignore the potential bias term and yield

$$\sqrt{T}(\check{\beta} - \beta) \longrightarrow_D \mathcal{N}(0, \Omega). \tag{3.88}$$

■

Proof of Lemma 3.3.2 and Theorem 3.3.3. We focus on the difference between $\check{\Sigma}_T$ and $\widehat{\Sigma}_T$. Write

$$\begin{aligned}
\widehat{\Sigma}_T &= \frac{1}{T}\sum_{t=1}^{T} \widehat{x}_t \widehat{x}_t' = \frac{1}{T}\sum_{t=1}^{T} \left(\widehat{x}_t - \check{x}_t + \check{x}_t\right)\left(\widehat{x}_t - \check{x}_t + \check{x}_t\right)' \\
&= \frac{1}{T}\sum_{t=1}^{T} (\widehat{x}_t - \check{x}_t)(\widehat{x}_t - \check{x}_t)' + \frac{1}{T}\sum_{t=1}^{T} (\widehat{x}_t - \check{x}_t)\check{x}_t' \\
&\quad + \frac{1}{T}\sum_{t=1}^{T} \check{x}_t(\widehat{x}_t - \check{x}_t)' + \frac{1}{T}\sum_{t=1}^{T} \check{x}_t \check{x}_t',
\end{aligned} \tag{3.89}$$

where the last term is $\check{\Sigma}_T$. Therefore,

$$\begin{aligned}\widehat{\Sigma}_T - \check{\Sigma}_T &= \frac{1}{T}\sum_{t=1}^{T}(\widehat{x}_t - \check{x}_t)(\widehat{x}_t - \check{x}_t)' + \frac{1}{T}\sum_{t=1}^{T}(\widehat{x}_t - \check{x}_t)\check{x}_t' \\ &\quad + \frac{1}{T}\sum_{t=1}^{T}\check{x}_t(\widehat{x}_t - \check{x}_t)' \\ &= \mathfrak{D}_{1T} + \mathfrak{D}_{2T} + \mathfrak{D}_{3T}, \end{aligned} \tag{3.90}$$

where $\mathfrak{D}_{1T} = T^{-1}\sum_{t=1}^{T}(\widehat{x}_t - \check{x}_t)(\widehat{x}_t - \check{x}_t)'$, $\mathfrak{D}_{2T} = T^{-1}\sum_{t=1}^{T}(\widehat{x}_t - \check{x}_t)\check{x}_t'$, and $\mathfrak{D}_{3T} = T^{-1}\sum_{t=1}^{T}\check{x}_t(\widehat{x}_t - \check{x}_t)'$. We control the difference by

$$\left|\left|\widehat{\Sigma}_T - \check{\Sigma}_T\right|\right| = ||\mathfrak{D}_{1T} + \mathfrak{D}_{2T} + \mathfrak{D}_{3T}|| \leq ||\mathfrak{D}_{1T}|| + ||\mathfrak{D}_{2T}|| + ||\mathfrak{D}_{3T}||. \tag{3.91}$$

Hence, $||\widehat{\Sigma}_T - \check{\Sigma}_T||$ converges to zeros in probability if $||\mathfrak{D}_l(T)|| \longrightarrow_P 0$ for $l = 1, 2, 3$. By Cauchy–Schwarz inequality,

$$\begin{aligned}||\mathfrak{D}_{2T}|| &= \left|\left|\frac{1}{T}\sum_{t=1}^{T}(\widehat{x}_t - \check{x}_t)\check{x}_t'\right|\right| \\ &\leq \left|\left|\frac{1}{T}\sum_{t=1}^{T}(\widehat{x}_t - \check{x}_t)(\widehat{x}_t - \check{x}_t)'\right|\right|^{1/2}\left|\left|\frac{1}{T}\sum_{t=1}^{T}\check{x}_t\check{x}_t'\right|\right|^{1/2} \\ &= ||\mathfrak{D}_{1T}||^{1/2}||\check{\Sigma}_T||^{1/2}. \end{aligned} \tag{3.92}$$

By Lemma 3.8.4, $||\mathfrak{D}_{1T}|| = o_p(1)$, and we have shown that the limit of $||\check{\Sigma}_T||$ is finite, therefore, $||\mathfrak{D}_{2T}|| = o_p(1)$. We can also show $||\mathfrak{D}_{3T}|| = o_p(1)$ using the same method. Thus, $||\widehat{\Sigma}_T - \check{\Sigma}_T|| \to_P 0$ as $T \to \infty$. Therefore, we complete the proof of Lemma 3.3.2 and Theorem 3.3.3. ■

Proof of Lemma 3.3.4 and Theorem 4.3.2. Note that

$$\sqrt{T}(\widehat{\beta}-\beta)=\sqrt{T}(\widehat{\beta}-\breve{\beta}+\breve{\beta}-\beta)=\sqrt{T}(\widehat{\beta}-\breve{\beta})+\sqrt{T}(\breve{\beta}-\beta), \tag{3.93}$$

where we have shown (3.88). Therefore, we can complete the proof once Lemma 3.3.4 is proved in which

$$\left|\left|\sqrt{T}(\widehat{\beta}-\breve{\beta})\right|\right|=o_p(1). \tag{3.94}$$

Note that $\widehat{Y}=\widehat{X}\beta+\widehat{\lambda}(V)+\widehat{U}$, $\breve{Y}=\breve{X}\beta+\breve{\lambda}(V)+\breve{U}$, $\widehat{e}=\widehat{\lambda}(\widehat{V})+\widehat{U}$, and $\breve{e}=\breve{\lambda}(V)+\breve{U}$. Here, $\widehat{Y}=(\widehat{y}_1,\ldots,\widehat{y}_T)'$, $\widehat{X}=(\widehat{x}_1,\ldots,\widehat{x}_T)'$, $\widehat{\lambda}(V)=(\widehat{\lambda}(v_1),\ldots,\widehat{\lambda}(v_T))'$, $\widehat{U}=(\widehat{u}_1,\ldots,\widehat{u}_T)'$, $\widehat{e}=(\widehat{e}_1,\ldots,\widehat{e}_T)'$ and $\widehat{y}_t=y_t-\widehat{\mathrm{E}}_h[y_t|\widehat{v}_t]$, $\widehat{x}_t=x_t-\widehat{\mathrm{E}}_h[x_t|\widehat{v}_t]$, $\widehat{\lambda}(v_t)=\lambda(v_t)-\widehat{\mathrm{E}}_h[\lambda(v_t)|\widehat{v}_t]$, $\widehat{u}_t=u_t-\widehat{\mathrm{E}}_h[u_t|\widehat{v}_t]$, $\widehat{e}_t=e_t-\widehat{\mathrm{E}}_h[e_t|\widehat{v}_t]$:

$$\begin{aligned}\widehat{\beta}-\breve{\beta}&=\widehat{\beta}-\beta-(\breve{\beta}-\beta)\\&=(\widehat{X}'\widehat{X})^{-1}(\widehat{X}'\widehat{Y})-\beta-((\breve{X}'\breve{X})^{-1}(\breve{X}'\breve{Y})-\beta)\\&=(\widehat{X}'\widehat{X})^{-1}(\widehat{X}'\widehat{e})-(\breve{X}'\breve{X})^{-1}(\breve{X}'\breve{e})\\&=(\widehat{X}'\widehat{X})^{-1}(\widehat{X}'\widehat{e})-(\widehat{X}'\widehat{X})^{-1}(\breve{X}'\breve{e})+(\widehat{X}'\widehat{X})^{-1}(\breve{X}'\breve{e})\\&\quad-(\breve{X}'\breve{X})^{-1}(\breve{X}'\breve{e})\\&=(\widehat{X}'\widehat{X})^{-1}(\widehat{X}'\widehat{e}-\breve{X}'\breve{e})+\left((\widehat{X}'\widehat{X})^{-1}-(\breve{X}'\breve{X})^{-1}\right)(\breve{X}'\breve{e}).\end{aligned} \tag{3.95}$$

Then our goal is to show that the norm of the following equation is $o_p(1)$:

$$\begin{aligned}\sqrt{T}\left(\widehat{\beta}-\breve{\beta}\right)&=\left(\frac{\widehat{X}'\widehat{X}}{T}\right)^{-1}\left(\frac{\widehat{X}'\widehat{e}-\breve{X}'\breve{e}}{\sqrt{T}}\right)\\&\quad+\left(\left(\frac{\widehat{X}'\widehat{X}}{T}\right)^{-1}-\left(\frac{\breve{X}'\breve{X}}{T}\right)^{-1}\right)\left(\frac{\breve{X}'\breve{e}}{\sqrt{T}}\right)\\&\triangleq DB_1+DB_2.\end{aligned} \tag{3.96}$$

We first examine DB_1. By Lemma 3.8.5, $\left|\left|(\widehat{X}'\widehat{e}-\breve{X}'\breve{e})/\sqrt{T}\right|\right|=o_p(1)$, and as shown in the first Theorem, $\widehat{\Sigma}_T=\frac{\widehat{X}'\widehat{X}}{T}\to_P\Sigma_g$, where

Σ_g is positive definite, hence $\widehat{\Sigma}_T^{-1} \to_P \Sigma_g^{-1}$. Then we have

$$DB_1 = O_p(1)O_p(1) = O_p(1). \tag{3.97}$$

We then examine DB_2. By the second step in the proofs of Theorem 3.3.3, $||\widehat{\Sigma}_T - \check{\Sigma}_T|| \to_P 0$, therefore,

$$\begin{aligned}\left(\frac{\widehat{X}'\widehat{X}}{T}\right)^{-1} - \left(\frac{\check{X}'\check{X}}{T}\right)^{-1} &= \left(\frac{\widehat{X}'\widehat{X}}{T}\right)^{-1}\left[\frac{\check{X}'\check{X}}{T} - \frac{\widehat{X}'\widehat{X}}{T}\right]\left(\frac{\check{X}'\check{X}}{T}\right)^{-1} \\ &= o_p(1). \end{aligned} \tag{3.98}$$

Meanwhile, in the first step, we have shown that $\check{X}'\check{e}/\sqrt{T} = O_p(1)$. Therefore,

$$DB_2 = O_p(1). \tag{3.99}$$

To summarize, equations (3.97) and (3.99) imply (3.94). Then, together with (3.88), we are able to complete the proof. ■

3.9 Bibliographical Notes

In this chapter, we have employed the conventional estimation method for the semiparametric partially linear model of equation (3.7) with different assumptions and settings for the regressors. In fact, the semiparametric regression model has been extensively studied in the literature. A primary introduction of semiparametric time series models as well as some useful examples are discussed in the works of Härdle *et al.* (2000), Fan and Yao (2003), Yatchew (2003), Härdle *et al.* (2004), Gao (2007), and Horowitz (2012).

Recent developments on non- and semi-parametric time series models as well as their applications include the works of Jensen and Maheu (2013), Gao and Phillips (2013), Sun *et al.* (2013), Saart *et al.* (2014), Chen *et al.* (2014), Li *et al.* (2015), Chen (2015), Boneva *et al.* (2015), Cai *et al.* (2015), Dong *et al.* (2015), Zhang *et al.* (2016), Jenish (2016), Fan and Liu (2016), Cai *et al.* (2018), Chen *et al.* (2018), Angrist *et al.* (2018), Patton *et al.* (2019), Hallin *et al.* (2022), Catania and Luati (2023), and the references therein.

Chapter 4

Time Series Regressions with Strong Trends

> The study of trends brings together empirical-quantitative and theory-quantitative aspects of modeling and has been empowered by that synergy.
>
> — Phillips (2001)

It is widely recognized that many empirical time series data, such as macroeconomic aggregates, financial prices, and climatic observations, exhibit dominant trend-like and random wandering behavior. This feature is often apparent by the first glance at their time series plots. However, when trending regressors are included in econometric models, we lack appropriate methods to provide a quantitative measure of the strength of trends and the statistical consequences for econometric regressions.

In a general trending regression model, the strength of the trend in the regressors undoubtedly affects the estimation and inference of the underlying regression relationship. In particular, when there exists the problem of endogeneity, the strength of the trend may determine the unbiasedness and consistency of, for example, the simple OLS estimators. In other words, the consequences of the endogeneity issue and the technical methods to address it depend heavily on the magnitude of trends, whether they are stochastic or deterministic.

This chapter investigates trending regression models with endogeneity when the regressors display strong trends that have dominant impacts on the entire series. To fix the endogeneity bias in the simple OLS estimators, we propose a bias-correction method that is particularly useful when the trends in the regressors are strong. In contrast to the previous chapter, where the focus was on weak trends in the regressors, we emphasize that the bias-correction method is only useful when the trending magnitudes of the regressors are greater than one. Both simulated and empirical examples are given to show the performance of the bias-correction method.

4.1 Model Specification

For a trending time series, we define an order of magnitude for its trend component, which measures how fast the time series grows to infinity with its sample size. This order of magnitude also indicates its strengths against stationary disturbances, which is similar to the concept of signal-to-noise ratio if the trend is the major component we are interested in. We assume that the data-generating process of a trending time series x_t is structured in an additive form as

$$x_t = g_t + v_t,$$

for $t = 1, 2, \ldots, T$, where g_t is the trend component and v_t is the remainder sequence of stationary disturbances with zero mean and finite variance. We denote by d the order of trending magnitude, which is defined based on the sum of squared g_t. Specifically, we assume that a trend component g_t has trending strength d (for some $d \geq 1$) if

$$\frac{1}{T^d} \sum_{t=1}^{T} g_t^2 = O_p(1), \tag{4.1}$$

while for all $d' < d$

$$\frac{1}{T^{d'}} \sum_{t=1}^{T} g_t^2 \to \infty, \tag{4.2}$$

when $T \to \infty$. Some typical examples are $d = 1$ for a weak trend $g_t = g(\tau_t)$, where $\tau_t = t/T$ and $g(\cdot)$ is a bounded function defined on [0, 1]; $d = 2$ when $g_t = x_{t-1}$ is a stochastic trend from a pure random walk process $x_t = x_{t-1} + v_t$, where v_t is a WN process with zero mean and finite variance; $d = 3$ when a simple linear time trend is in x_t such as $g_t = 1+0.01t$. Furthermore, d can also be a fractional number when x_t is a sequence of fractionally integrated time series or contains a deterministic trend of $(d-1)/2$th order time polynomial for any real value $d \geq 1$.

We define g_t as a weak trend when $d = 1$, and otherwise a strong trend when $d > 1$. In the previous chapter, we discussed the linear regression model in which all the trends are weak ($d = 1$). As the weak trend does not dominate the entire time series, the endogenous correlation between the regressor and the regression error induces bias and inconsistency in, for example, the OLS estimator. To deal with the problem of endogeneity, we propose a nonparametric control function approach to obtain consistent estimators for the regression coefficients in the trending regression models with weak trends.

In this chapter, we consider regressors with strong trends. Note that in this case, we have $d > 1$, and the sum of squared x_t has the same order of magnitude as the sum of squared g_t. Specifically,

$$\begin{aligned}\frac{1}{T^d}\sum_{t=1}^{T} x_t^2 &= \underbrace{\frac{1}{T^d}\sum_{t=1}^{T} g_t^2}_{S_{gg}} + \underbrace{\frac{1}{T^d}\sum_{t=1}^{T} v_t^2}_{S_{vv}} + \underbrace{2\frac{1}{T^d}\sum_{t=1}^{T} g_t v_t}_{S_{gv}} \\ &\sim \frac{1}{T^d}\sum_{t=1}^{T} g_t^2, \end{aligned} \tag{4.3}$$

where $S_{vv} = T^{-d}\sum_{t=1}^{T} v_t^2 = O_p(T^{-(d-1)})$ since v_t has finite variance, and by the Cauchy–Schwarz inequality,

$$\begin{aligned}|S_{gv}| = \left|\frac{1}{T^d}\sum_{t=1}^{T} g_t v_t\right| &\leq \frac{1}{T^{(d-1)/2}}\sqrt{\left(\frac{1}{T^d}\sum_{t=1}^{T} g_t^2\right)\left(\frac{1}{T}\sum_{t=1}^{T} v_t^2\right)} \\ &= O_p\left(\frac{1}{T^{(d-1)/2}}\right), \end{aligned} \tag{4.4}$$

so that both S_{vv} and S_{gv} are negligible terms for $d > 1$ as $T \to \infty$. Thus, the limits of $\sum_{t=1}^{T} x_t^2$ and $\sum_{t=1}^{T} g_t^2$ have the same order indicating that the strong trend g_t dominates the time series of x_t and the stationary component v_t plays a secondary role when the sample size is large.

In this chapter, we revisit the linear regression model

$$y_t = x_t'\beta + e_t, \tag{4.5}$$

$$x_{it} = g_{it} + v_{it}, \tag{4.6}$$

for $i = 1, 2, \ldots, k$, $t = 1, 2, \ldots, T$, where $x_t = (1, x_{1t}, \ldots, x_{kt})'$ includes the intercept term as its first element, and we also assume that all the trending regressors $(x_{1t}, \ldots, x_{kt})'$ contain strong trends, i.e., $g_t = (g_{1t}, \ldots, g_{kt})'$ that $d_i > 1$ denotes the trending strength of g_{it}, respectively, for all $i = 1, 2, \ldots, k$. As usual, the error terms $(e_t, v_t')'$ are stationary with zero mean and positive definite variance–covariance matrix, where $v_t = (v_{1t}, \ldots, v_{kt})'$. The problem of endogeneity arises when e_t is correlated with at least one of the elements in v_t.

When the trends $(g_{1t}, \ldots, g_{kt})'$ are strong, they play a dominant role over the stationary components, where the issue of endogeneity occurs. As a result, we show in detail later that the simple OLS estimator of β turns out to be consistent when T tends to infinity. However, the limit distribution of the OLS estimator may not be centered around zero. Therefore, the OLS method provides consistent estimates, but the inference conclusions are likely to be unreliable due to the problem of endogeneity.

4.2 Assumptions

We emphasize that only regression models in which all the regressors exhibit strong trends are considered. As one could always apply the nonparametric control function approach, we do not discuss the mixture scenario for which there exist weak and strong trends at the same time. Therefore, all the time trends play dominant roles over the stationary disturbances. In order to carry out the asymptotic results, we regulate the nonlinear and nonparametric time trends by introducing some matrix notations and necessary assumptions as follows.

In matrix notation, our regression model is

$$\boldsymbol{y} = \boldsymbol{X}\boldsymbol{\beta} + \boldsymbol{e}, \tag{4.7}$$

$$\boldsymbol{X} = \boldsymbol{G} + \boldsymbol{v}, \tag{4.8}$$

where $\boldsymbol{y} = (y_1, \ldots, y_T)'$, $\boldsymbol{e} = (e_1, \ldots, e_T)'$, $\beta = (\beta_0, \beta_1, \ldots, \beta_k)'$, and $\boldsymbol{X}$ is a $T \times (k+1)$ matrix for the regressors

$$\boldsymbol{X} = \begin{pmatrix} 1 & x_{11} & \ldots & x_{k1} \\ 1 & x_{12} & \ldots & x_{k2} \\ \ldots & \ldots & \ldots & \ldots \\ 1 & x_{1T} & \ldots & x_{kT} \end{pmatrix},$$

where

$$\boldsymbol{G} = \begin{pmatrix} 1 & g_{11} & \cdots & g_{k1} \\ 1 & g_{12} & \cdots & g_{k2} \\ \ldots & \ldots & \ldots & \ldots \\ 1 & g_{1T} & \cdots & g_{kT} \end{pmatrix},$$

and

$$\boldsymbol{v} = \begin{pmatrix} 0 & v_{11} & \ldots & v_{k1} \\ 0 & v_{12} & \ldots & v_{k2} \\ \ldots & \ldots & \ldots & \ldots \\ 0 & v_{1T} & \ldots & v_{kT} \end{pmatrix},$$

for $i = 1, 2, \ldots, k$ and $t = 1, 2, \ldots, T$.

Assumption 4.2.1. Assume that there exists a $(k+1)$-dimensional diagonal matrix $D = diag(T^{1/2}, T^{d_1/2}, \ldots, T^{d_k/2})$ such that $d_i > 1$, for $i = 1, 2, \ldots, k$, are the orders of magnitude of the corresponding trend component g_{it} such that as $T \to \infty$:

$$D^{-1}\boldsymbol{G}'\boldsymbol{G}D^{-1} \longrightarrow_P Q, \tag{4.9}$$

where Q is a $(k+1)$-dimensional positive definite matrix.

Remark 4.2.1. This assumption first rules out weak trends in x_t since parameters d_i are assumed to be greater than one for $i = 1, 2, \ldots, k$. Matrix Q is a positive definite matrix with full rank, therefore, multicollinearity is impossible and none of the time trends can be represented by linear functions of the others. In other words, this condition rules out the possibility of common trends within the set of $(x_{1t}, \ldots, x_{kt})'$.

Specifically, the first row and column of matrix $D^{-1}\boldsymbol{G}'\boldsymbol{G}D^{-1}$ are the rescaled sample means of $(1, g_{i1}, g_{i2}, \ldots, g_{iT})'$. We assume that as $T \to \infty$, such sample means satisfy $T^{-(d_i+1)/2}\sum_{t=1}^{T} g_{it} \to Q_{1,i+1}$, and for the other elements in $D^{-1}\boldsymbol{G}'\boldsymbol{G}D^{-1}$,

$$\frac{1}{T^{d_{ij}}}\sum_{t=1}^{T} g_{it}g_{jt} \longrightarrow Q_{i+1,j+1}, \tag{4.10}$$

for $i, j = 1, 2, \ldots, k$, where $d_{ij} = (d_i + d_j)/2$. In particular, when $i = j$, the condition is exactly how we have defined the order of magnitude of a trend component g_{it} as

$$\frac{1}{T^{d_i}}\sum_{t=1}^{T} g_{it}^2 \longrightarrow Q_{i+1,i+1}, \tag{4.11}$$

for $i = 1, 2, \ldots, k$.

Assumption 4.2.2. Let $(\varepsilon_t, \boldsymbol{\eta}_t')' = (\varepsilon_t, \eta_{1t}, \ldots, \eta_{kt})'$ be a $k+1$ vector of i.i.d. innovations with mean zero and $\mathrm{E}[\varepsilon_1^2] = \sigma_1^2$, $(\theta_1, \ldots, \theta_k)' = \boldsymbol{\Theta}$ are the covariances between ε_t and η_{it} that $\mathrm{Cov}(\varepsilon_t, \eta_{it}) = \theta_i$ and $\boldsymbol{\Sigma}_{22}$ is the $k \times k$ variance–covariance matrix of $\boldsymbol{\eta}_t = (\eta_{1t}, \ldots, \eta_{kt})'$ with σ_{ij} being the element at the ith row and jth column, i.e., $\sigma_{ij} = \mathrm{E}[\eta_{it}\eta_{jt}]$. Meanwhile, we assume that $\mathrm{E}[\varepsilon_t^4] < \infty$, $\mathrm{E}[\eta_{it}^4] < \infty$, and $\mathrm{E}[\varepsilon_t^4\eta_{it}^4] < \infty$, for $i, j = 1, 2, \ldots, k$.

Remark 4.2.2. This assumption defines an i.i.d. sequence of innovation vectors with zero mean and variance–covariance matrix:

$$Var\left((\varepsilon_t, \boldsymbol{\eta}_t')'\right) = \begin{pmatrix} \sigma_1^2 & \boldsymbol{\Theta}' \\ \boldsymbol{\Theta} & \Sigma_{22} \end{pmatrix}. \tag{4.12}$$

Assumption 4.2.3. The error terms are defined as linear processes with respect to the sequences of innovations $(\varepsilon_t, \boldsymbol{\eta}_t')'$ in Assumption (4.2.2). Specifically,

$$e_t = \sum_{s=0}^{\infty} \phi_s \varepsilon_{t-s} \triangleq \Phi(L)\varepsilon_t, \tag{4.13}$$

$$v_t = \sum_{s=0}^{\infty} \boldsymbol{\psi}_s \boldsymbol{\eta}_{t-s} \triangleq \boldsymbol{\Psi}(L)\boldsymbol{\eta}_t, \tag{4.14}$$

where

$$\Phi(L) = \sum_{s=0}^{\infty} \phi_s L^s, \tag{4.15}$$

$$\boldsymbol{\Psi}(L) = \sum_{s=0}^{\infty} \boldsymbol{\psi}_s L^s, \tag{4.16}$$

in which $\boldsymbol{\psi}_s = \text{diag}(\psi_{s,1}, \ldots, \psi_{s,k})$ is a $k \times k$-diagonal matrix and L is the *lag-operator*, for example, $L^i x_t = x_{t-i}$. The coefficients satisfy $\sum_{s=0}^{\infty} \phi_s^2 < \infty$ and $\sum_{s=0}^{\infty} \psi_{s,i}^2 < \infty$, for $i = 1, 2, \ldots, k$.

To be adaptive with the matrix notations, we let $\boldsymbol{v}_t = (0, v_t')'$ for $t = 1, 2, \ldots, T$, where $v_t = (v_{1t}, v_{2t}, \ldots, v_{kt})'$. To explore and understand the structure of the error terms, we expand the expression of v_t in equation (4.14). As $\boldsymbol{\psi}_s$ is a diagonal matrix, we can write the elements of v_t as

$$v_{it} = \Psi_i(L)\eta_{i,t} = \sum_{s=0}^{\infty} \psi_{s,i}\eta_{i,t-s}, \tag{4.17}$$

for $i = 1, 2, \ldots, k$. Meanwhile, we define

$$f_{i,q}(L) = \sum_{s=0}^{\infty} \phi_s \psi_{s+q,i} L^s, \tag{4.18}$$

$$m_{i,q}(L) = \sum_{s=0}^{\infty} \phi_{s+q} \psi_{s,i} L^s, \tag{4.19}$$

and by the Beveridge–Nelson decomposition, we have

$$\Phi(L) = \Phi(1) - (1 - L)\widetilde{\Phi}(L), \tag{4.20}$$

$$\Psi_i(L) = \Psi(1) - (1 - L)\widetilde{\Psi}_i(L), \tag{4.21}$$

$$f_{i,q}(L) = f_{i,q}(1) - (1 - L)\widetilde{f}_{i,q}(L), \tag{4.22}$$

$$m_{i,q}(L) = m_{i,q}(1) - (1 - L)\widetilde{m}_{i,q}(L), \tag{4.23}$$

for $i = 1, 2, \ldots, k$, where $\widetilde{\Phi}(L) = \sum_{s=0}^{\infty} \widetilde{\phi}_s L^s$, $\widetilde{\Psi}_i(L) = \sum_{s=0}^{\infty} \widetilde{\psi}_{s,i} L^s$, $\widetilde{f}_{i,q}(L) = \sum_{s=0}^{\infty} \widetilde{f}_{i,qs} L^s$, and $\widetilde{m}_{i,q}(L) = \sum_{s=0}^{\infty} \widetilde{m}_{i,qs} L^s$, in which the new coefficients are $\widetilde{\phi}_s = \sum_{p=s+1}^{\infty} \phi_p$, $\widetilde{\psi}_{s,i} = \sum_{p=s+1}^{\infty} \psi_{p,i}$, $\widetilde{f}_{i,qs} = \sum_{p=s+1}^{\infty} \phi_p \psi_{p+q,i}$, and $\widetilde{m}_{i,qs} = \sum_{p=s+1}^{\infty} \phi_{p+q} \psi_{p,i}$.

Remark 4.2.3. In Assumption 4.2.3, both the error terms e_t and v_t follow linear processes whose innovations ε_t and $\boldsymbol{\eta}_t$ are correlated by following a joint probability distribution in Assumption 4.2.2. In fact, it is a special case of the linear process $u_t = (e_t, v_t')$ that $u_t = \sum_{s=0}^{\infty} \gamma_s \mu_{t-s} \triangleq \Gamma(L)\mu_t$, where μ_t is a WN process with mean $\mathbf{0}$ and variance–covariance matrix Σ_μ for $t = 1, 2, \ldots, T$, and $\Gamma(L) = \sum_{s=0}^{\infty} \gamma_s L^s$.

When $\boldsymbol{\Theta} \neq 0$, the innovations ε_t and η_t are correlated and therefore, the error terms v_t and e_t are also correlated. As a result, the correlation causes the problem of endogeneity in the regression.

We define $\widehat{Q} = D^{-1}\boldsymbol{X}'\boldsymbol{X}D^{-1}$. The strong trend condition $d_i > 1$ indicates that each regressor x_{it} is dominated by g_{it}, whose squared sum has a higher order of magnitude than that of the stationary disturbance v_{it}. Hence, we have the following theorem.

Theorem 4.2.1. *Under Assumptions* 4.2.1–4.2.3, *as* $T \to \infty$, *we have*

$$\widehat{Q} = D^{-1}\boldsymbol{X}'\boldsymbol{X}D^{-1} \longrightarrow_P Q, \tag{4.24}$$

where Q is the positive definite matrix in Assumption 4.2.1.

Remark 4.2.4. As the regressors are all dominated by their trend components, Theorem 4.2.1 shows that the variation in the time trend plays a central role in the identification and estimation of the regression coefficient β. That is, the probability limit of Q must be positive definite, otherwise it is an interesting question to discuss the identification of, for example, $y_t = \beta_0 + x_{1t}\beta_1 + x_{2t}\beta_2 + e_t$, where the regressors contain the same form of strong linear trends that $x_{it} = a_i + b_i t + v_{it}$, for $b_i \neq 0$, $i = 1, 2$, and $t = 1, 2, \ldots, T$. In particular, the corresponding Q becomes nonsingular when T goes to infinity, and proper identification of the regression coefficients may rely on the stationary components of x_{1t} and x_{2t}.

4.3 The OLS and Bias-corrected Estimator

Before introducing the techniques to deal with the problem of endogeneity in the regression equation (4.5), we first investigate the asymptotic properties of its simple OLS estimator under the condition that all the regressors x_{it} contain time trends for $i = 1, 2, \ldots, k$. Recall that the OLS estimator is defined as

$$\widehat{\beta}_{\text{ols}} = (\boldsymbol{X}'\boldsymbol{X})^{-1}\boldsymbol{X}'\boldsymbol{y} = \left(\sum_{t=1}^{T} x_t x_t'\right)^{-1} \sum_{t=1}^{T} x_t y_t. \tag{4.25}$$

In Theorem 4.2.1, we have shown that $D^{-1}\boldsymbol{X}'\boldsymbol{X}D^{-1} \longrightarrow_P Q$ as $T \to \infty$, in which Q is a positive definite matrix. Therefore, $\boldsymbol{X}'\boldsymbol{X}$ is invertible in finite sample and $\widehat{\beta}_{\text{ols}}$ is legitimately defined. Furthermore, by simple transformation and proper rescaling, we have

$$D\left(\widehat{\beta}_{\text{ols}} - \beta\right) = \left(D^{-1}\boldsymbol{X}'\boldsymbol{X}D^{-1}\right)^{-1}\left(D^{-1}\boldsymbol{X}'\boldsymbol{e}\right). \tag{4.26}$$

Given that the right-hand side of equation (4.26) is $O_p(1)$, it is apparent that the convergence rate of the coefficient element $\widehat{\beta}_{ols,i}$ is T^{-d_i}, and this is highly dependent on the strength of g_{it}, for $i = 1, 2, \ldots, k$, respectively.

When the problem of endogeneity exists in the regression equation, the conditional expectation of $\boldsymbol{e}$ given $\boldsymbol{X}$ is nonzero, causing $D^{-1}\boldsymbol{X}'\boldsymbol{e}$ to diverge to infinity. As a consequence, a bias term has to be considered in the asymptotic results.

Theorem 4.3.1. *Under Assumptions* 4.2.1–4.2.3, *as* $T \to \infty$,

$$D\left(\widehat{\beta}_{\text{ols}} - \beta - D^{-1}\widehat{Q}^{-1}D^{-1}T\tilde{b}\right) \longrightarrow_D \mathcal{N}(0, \Omega^*), \tag{4.27}$$

where $\tilde{b} = T^{-1}\sum_{t=1}^{T} \boldsymbol{v}_t' e_t$ *with limit* $b = (0, b_1, \ldots, b_k)'$, $\Omega^* = Q^{-1}\Omega Q^{-1}$ *in which* Ω *is the asymptotic variance–covariance matrix of* $D^{-1}\boldsymbol{G}'\boldsymbol{e}$.

Remark 4.3.1. The probability limit of $\widehat{b}$ is $b = (0, (\sum_{j=0}^{\infty} \phi_j \boldsymbol{\Psi}_j \boldsymbol{\Theta})')$, where $\phi_j, \boldsymbol{\Psi}_j$, and $\boldsymbol{\Theta}$ are from Assumption 4.2.3. This indicates that when $\boldsymbol{\Theta}$ is a nonzero vector, the endogenous correlation between the error terms e_t and v_t induces a bias term in the asymptotic result of the simple OLS estimator. Since we have assumed that all the trends g_{it} are strong that dominate the time series, $d_i > 1$ for all i ensures that the bias term $D^{-1}\widehat{Q}^{-1}D^{-1}T\tilde{b}$ diminishes to zero when $T \to \infty$. Therefore, the OLS method provides a consistent estimator for the regression coefficients. Meanwhile, the asymptotic variance of Ω has the form of $\Phi(1)^2\sigma_1^2 Q$.

If, however, g_{1t} is a weak trend while $g_{2t}, \ldots, g_{kt}$ are strong trends for $k > 1$, then the vector of bias term $D^{-1}\widehat{Q}^{-1}D^{-1}T\tilde{b}$ does not diminish to zero with sample size T, and the OLS estimators for $\beta_2, \ldots, \beta_k$ are also inconsistent if some of the off-diagonal elements of $\widehat{Q}$ are nonzero.

To investigate the bias and the asymptotic distribution under different orders of trends, we provide a corollary for the univariate regression.

Corollary 4.3.1. *Under Assumptions* 4.2.1–4.2.3, *let* $k = 1$. *As* $T \to \infty$, *we have*

$$D\left(\widehat{\beta}_{\text{ols}} - \beta - B_T\right) \to_D \mathcal{N}(0, \Omega^*), \tag{4.28}$$

where $D = \text{diag}(\sqrt{T}, \sqrt{T^d})$, $B_T = diag(T^{(1-d)/2}, T^{1-d})q_{,2}^{inv}\tilde{b}_1$ *is the bias term in which* $\tilde{b}_1 = T^{-1}\sum_{t=1}^{T} e_t v_t$, *and* $q_{,2}^{inv} = (q_{12}^{inv}, q_{22}^{inv})'$ *is the second column of matrix* $\widehat{Q}^{-1}$.

We find that the intercept $\widehat{\beta}_{\text{ols},0}$ has a bias of order $O_p(T^{(1-d)/2})$ and the slope coefficient $\widehat{\beta}_{\text{ols},1}$ has a bias of order $O_p(T^{1-d})$. As $d > 1$, both bias terms converge to zero with $T \to \infty$, indicating that the OLS estimator is consistent. However, the performance of the OLS estimator differs for different values of d.

Corollary 4.3.2. *Under Assumptions* 4.2.1–4.2.3, *let* $\mathbf{\Theta} \neq 0$ *and* $k = 1$. *For the OLS estimator of the coefficient, we have the following results:*

- *When* $1 < d < 2$, *the OLS estimators are consistent but biased. Since* DB_T *diverges to infinity, the existence of bias renders the application of regular statistical inference though the bias term* B_T *diminishes to zero with sample size* T.

- *When* $d = 2$*, the estimator is super-consistent. However, the OLS estimator converges to a distribution*

$$D(\widehat{\beta}_{\text{ols}} - \beta) \longrightarrow_D \mathcal{N}(B, \Omega^*), \tag{4.29}$$

that is not centered around zero, where $B = q_{,2}^{inv} b_1$ *is a nonzero* 2×1 *constant vector when there exists endogeneity.*
- *For* $d > 2$*, the potential bias term satisfies* $DB_T = o_P(1)$*, which is negligible compared to the limit distribution that is always regarded as* $O_P(1)$*. Therefore, the simple OLS estimator could be treated as unbiased and consistent when the sample size tends to infinity:*

$$D(\widehat{\beta}_{\text{ols}} - \beta) \longrightarrow_D \mathcal{N}(0, \Omega^*). \tag{4.30}$$

Since the simple OLS estimator is always consistent when $d_i > 1$, the bias in the OLS estimator can therefore be estimated consistently. We propose a bias-corrected estimator as

$$\widehat{\beta}_{\text{bc}} = \widehat{\beta}_{\text{ols}} - \widehat{b}, \tag{4.31}$$

where $\widehat{b} = \left(\sum_{t=1}^{T} x_t x_t'\right)^{-1} \sum_{t=1}^{T} \widehat{v}_t \widehat{e}_t$ estimates the bias term, in which $\widehat{e}_t = y_t - x_t' \widehat{\beta}_{\text{ols}}$, $\widehat{\boldsymbol{v}}_t = (0, \widehat{v}_{1t}, \ldots, \widehat{v}_{kt})'$, $\widehat{v}_{it} = x_{it} - \widehat{g}_{it}$, and $\widehat{g}_{it}$ is the nonparametric estimates of the time trend in x_{it} for $i = 1, 2, \ldots, k$.

Remark 4.3.2. A major advantage of this bias-correction method is that we do not need to know the exact value of d_i for $i = 1, 2, \ldots, k$ when we employ the bias-correction procedure for $\widehat{\beta}_{\text{ols}}$. The main reason is that d_i is not involved in the estimator of the bias term in equation (4.31).

Theorem 4.3.2. *Under Assumptions* 4.2.1–4.2.3*, the bias-corrected estimator is unbiased and consistent that as* $T \to \infty$*,*

$$D\left(\widehat{\beta}_{\text{bc}} - \beta\right) \longrightarrow_D \mathcal{N}(0, \Omega^*), \tag{4.32}$$

where $\Omega^* = Q^{-1} \Omega Q^{-1}$ *are defined the same as in Theorem* 4.3.1.

Remark 4.3.3. Given that Assumptions 4.2.1–4.2.3 hold, we are able to simplify Ω, the variance of $D^{-1} \boldsymbol{G}' \boldsymbol{e}$, to $\Phi(1)^2 \sigma_1^2 Q$. Then, the variance of $\widehat{\beta}_{\text{bc}}$ can be estimated by $\Phi(1)^2 \widehat{\sigma}_1^2 (\boldsymbol{X}' \boldsymbol{X})$ provided that

$\widehat{\sigma}_1^2$ is a consistent estimator of σ_1^2. We emphasize that the trending magnitude parameters d_i are also not involved in the estimation of the variance–covariance matrix of $\widehat{\beta}_{\rm bc}$.

The availability of the bias-correction method depends on the consistency of the first-stage OLS estimator, and this is the same as the bias-correction method used for dealing with the problem of endogeneity in the cointegration regression by Phillips and Hansen (1990).

In theory, the bias-corrected estimator $\widehat{\beta}_{\rm bc}$ is closer to β, the true value of the regression coefficients, than the first-stage OLS estimator $\widehat{\beta}_{\rm ols}$ because the endogeneity bias has been reduced to some extent. Therefore, the corresponding residual sequence of $\widehat{e}_t^* = y_t - x_t\widehat{\beta}_{\rm bc}$ must give a better approximation of the true errors e_t than the OLS residuals $\widehat{e}_t$, for $t = 1, 2, ..., T$. As a result, the estimated bias $\widehat{b}^* = \left(\sum_{t=1}^T x_t x_t'\right)^{-1} \sum_{t=1}^T \widehat{v}_t \widehat{e}_t^*$ must be more accurate than $\widehat{b}$ which uses less accurate approximations of e_t. Thus, we suggest that to obtain accurate estimates of the coefficients, the bias-correction procedure could be iterated several times until convergence. To summarize, the implementation steps are listed as follows:

(1) Estimate g_t in x_t using the nonparametric kernel estimation method and compute the residuals as $\widehat{v}_t = x_t - \widehat{g}_t$.
(2) Compute the OLS estimator $\widehat{\beta}_{\rm ols}$, and obtain the OLS residuals $\widehat{e}_t = y_t - x_t'\widehat{\beta}_{\rm ols}$. Estimate the bias term by $\widehat{b} = \left(\sum_{t=1}^T x_t x_t'\right)^{-1} \sum_{t=1}^T \widehat{v}_t \widehat{e}_t$.
(3) Compute the bias-corrected estimator $\widehat{\beta}_{\rm bc} = \widehat{\beta}_{\rm ols} - \widehat{b}$.
(4) Compute the new residuals $\widehat{e}_t^* = y_t - x_t\widehat{\beta}_{\rm bc}$, and re-estimate the bias term by $\widehat{b}^* = \left(\sum_{t=1}^T x_t x_t'\right)^{-1} \sum_{t=1}^T \widehat{v}_t \widehat{e}_t^*$.
(5) Compute the bias-corrected estimator $\widehat{\beta}_{\rm bc} = \widehat{\beta}_{\rm ols} - \widehat{b}^*$.

Repeat Steps 4 and 5 sufficiently many times or until the convergence of $\widehat{\beta}_{\rm bc}$.

4.4 Estimation of the Trending Magnitude

Although we do not need to know the exact values of d_i in the bias-correction procedure, we sometimes need to approximate their values.

If parametric specifications are imposed on g_{it}, d_i could be evaluated in a way that ensures the theoretical convergence of $T^{-d_i}\sum_{t=1}^{T} g_{it}^2$. For example, $d_i = 3$ for linear time trends. If g_{it} are left as nonparametric forms, one way to estimate d_i is introduced as follows.

We have shown that the unobserved $\sum_{t=1}^{T} g_{it}^2$ has the same order of magnitude $O_p(T^{d_i})$ as $\sum_{t=1}^{T} x_{it}^2$, which is observable. We have

$$\sum_{t=1}^{T} x_{it}^2 = Q_i T^{d_i}. \tag{4.33}$$

We take the logarithms of both sides and yield

$$\ln\left(\sum_{t=1}^{T} x_{it}^2\right) = \ln(Q_i) + d_i \ln T. \tag{4.34}$$

Therefore,

$$d_i = \frac{\ln\left(\sum_{t=1}^{T} x_{it}^2\right)}{\ln T} + \frac{\ln(Q_i)}{\ln T}, \tag{4.35}$$

where the latter term $\ln(Q_i)/\ln T$ is $o_p(1)$. Since $1/\ln T \to 0$ as $T \to \infty$, we define a consistent estimator for d_i as

$$\widehat{d_i} = \frac{\ln\left(\sum_{t=1}^{T} x_{it}^2\right)}{\ln T}, \tag{4.36}$$

for $i = 1, 2, \ldots, k$.

Remark 4.4.1. The problem in the estimator of d_i is that $\ln T$ goes to infinity very slowly with T so that the second term in (4.35) can be quite large in finite sample, resulting in a relatively quite large bias in $\widehat{d_i}$.

4.5 Simulated Example

In this section, we investigate the behaviors of the simple OLS estimator and the bias-corrected estimator in the context of strong trending regression models with endogeneity. We also show the improvement in statistical inference when the endogeneity bias has been properly adjusted. We consider a univariate regression model formulated as

$$y_t = \alpha + x_t\beta + e_t, \tag{4.37}$$

$$x_t = g_t + v_t, \tag{4.38}$$

where the coefficients are $\alpha = 0.8$, $\beta = 0.5$. Both error terms e_t and v_t follow AR(1) processes as $e_t = 0.2e_{t-1} + \varepsilon_t$ and $v_t = 0.2v_{t-1} + \eta_t$, where $(\varepsilon_t, \eta_t)' \overset{i.i.d.}{\sim} \mathcal{N}(0, \Sigma)$ and

$$\Sigma = \begin{pmatrix} 1 & 0.5 \\ 0.5 & 1 \end{pmatrix}, \tag{4.39}$$

where the off-diagonal element is nonzero. Hence, e_t and v_t are correlated, and so are e_t and x_t. This causes the problem of endogeneity in the linear regression model of y_t over x_t. To explore the impact of endogeneity and the performance of the bias-correction approach, we consider different trending scenarios in x_t. Specifically, for $t = 1, 2, \ldots, T$, we consider three trend specifications as

- Example 1: $g_t = 0.7\sqrt[3]{t}$,
- Example 2: $g_t = 0.3\sqrt{t}$,
- Example 3: $g_t = 0.01t$.

The first example corresponds to the case that $1 < d < 2$, a relatively weak one among the collection of strong trends. In particular, d equals 5/3 for this example where the trend term g_t goes to infinity with a diminishing speed, and the OLS estimator is biased but consistent. The second example demonstrates a deterministic trend with trending magnitude $d = 2$, which is stronger than that in Example 1, and the OLS estimator of the regression coefficient is consistent, but the limiting distribution of $T(\widehat{\beta}_{\text{ols}} - \beta)$ is not centered around zero as it has a nondiminishing bias caused by the problem of endogeneity. In the third example, the regressor contains a linear function of time with trending magnitude $d = 3$, which is the strongest among the three examples. The OLS estimator is unbiased and consistent in that it has a very strong trend that grows linearly with t.

With the sample size set as $T = 200, 400, 800$, and 1600, respectively, we simulate $\{x_t, y_t\}_{t=1}^T$ according to the data-generating processes we have designed. Then we compute the OLS estimator $\widehat{\beta}^{ols}$

and the bias-corrected estimators $\widehat{\beta}^{bc,1}$ and $\widehat{\beta}^{bc,iter}$, where the bias-correction procedure has been implemented for only once and iteratively for many times, respectively. We repeat the simulation process independently for $N_B = 5{,}000$ times and denote the estimates of β based on the OLS and bias-correction method as $\widehat{\beta}_j^{ols}$, $\widehat{\beta}_j^{bc,1}$, and $\widehat{\beta}_j^{bc,iter}$, respectively, for $j = 1, 2, \ldots, N_B$. Table 4.1 shows the biases, standard deviations, and the root mean squared errors for the estimators $\widehat{\beta}_p^{ols}$, $\widehat{\beta}_p^{bc,1}$, and $\widehat{\beta}_p^{bc,iter}$ under different sample sizes in Examples one to three. Some important findings are summarized as follows.

First, the performance of the OLS estimator depends on the strength of trends in the regressors. When the trend in x_t gets stronger from Examples 1–3, the endogeneity bias in the OLS estimator diminishes. Specifically, in Example 1, when the trend is not strong enough (although it belongs to the strong trend category and the OLS estimator is consistent in theory), the OLS estimator always contains a large bias that decreases extremely slowly in finite sample. In other words, gathering more data to increase sample size does not help eliminate the endogeneity bias in the OLS estimator effectively. However, when the trend in x_t gets stronger in Example 2 and even stronger in Example 3, a larger sample size results in a more precise estimator with a much reduced bias, due to the consistency of the OLS estimator.

Second, the bias-correction method does reduce bias for strong trending regression models. In particular, for the case when the bias-correction procedure is applied only once, the bias gets smaller in all examples, and therefore, the RMSEs get smaller after bias correction. The effects are more evident for the examples with stronger trends and larger sample sizes. We also find that when the trend is not sufficiently strong, or when the sample size is quite small, the estimator with bias correction applied only once still carries a non-negligible bias.

Third, iteration greatly improves the accuracy of the bias-corrected estimators, and this is a striking finding in Table 4.1. Their RMSEs reduce a lot compared with those when bias correction is applied only once. As we apply the bias-correction procedure based on newly obtained estimates of the coefficients iteratively, bias is further reduced by a large extent, indicating that the bias-corrected estimators are moving toward the true values of the coefficients.

Table 4.1. Simulation results for the strongly trending regression with endogeneity.

		OLS			Bias correction					
					Applied for only once			Applied with iteration		
	T	Bias	Sd	RMSE	Bias	Sd	RMSE	Bias	Sd	RMSE
Example 1	200	0.4827	0.0637	0.4869	0.4625	0.0682	0.4675	0.2941	0.1873	0.3487
	400	0.4647	0.0438	0.4668	0.4317	0.0485	0.4344	0.1781	0.1337	0.2227
	800	0.4457	0.0319	0.4469	0.3971	0.0362	0.3987	0.0909	0.0955	0.1318
	1600	0.4187	0.0218	0.4192	0.3493	0.0252	0.3502	0.0302	0.0605	0.0677
Example 2	200	0.3503	0.0572	0.3549	0.2465	0.0672	0.2555	0.0163	0.1239	0.1249
	400	0.2689	0.0384	0.2716	0.1446	0.0460	0.1518	0.0041	0.0669	0.0671
	800	0.1841	0.0218	0.1854	0.0669	0.0267	0.0721	0.0025	0.0334	0.0335
	1600	0.1140	0.0129	0.1148	0.0259	0.0152	0.0300	0.0013	0.0165	0.0166
Example 3	200	0.3787	0.0589	0.3833	0.2908	0.0699	0.2991	0.0313	0.1452	0.1486
	400	0.2197	0.0346	0.2225	0.0992	0.0416	0.1075	0.0059	0.0552	0.0555
	800	0.0817	0.0159	0.0832	0.0140	0.0180	0.0228	0.0009	0.0188	0.0188
	1600	0.0233	0.0065	0.0242	0.0013	0.0067	0.0068	0.0002	0.0067	0.0067

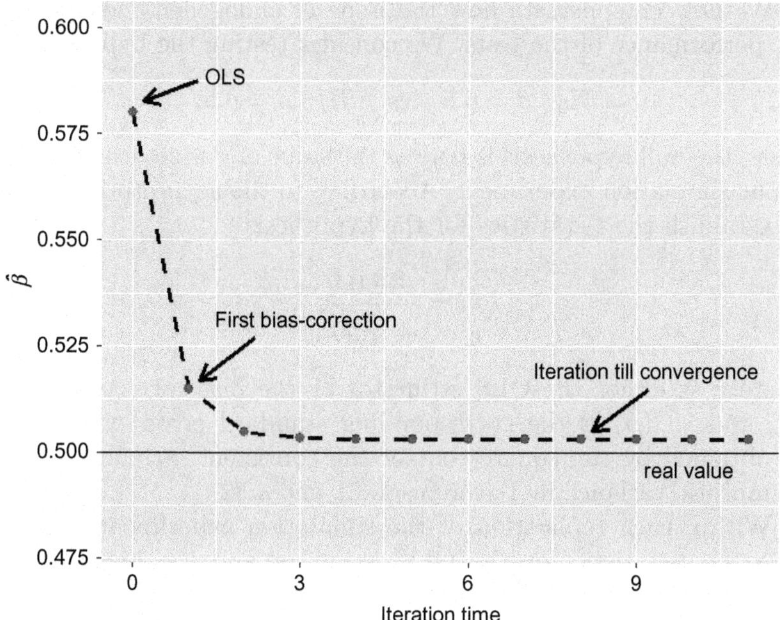

Figure 4.1. Iteration of the bias-correction procedure.

Figure 4.1 shows how the bias-corrected estimator approaches the true value of β through iteration.

To obtain accurate estimates of the regression coefficients, our simulation results in the above table suggest that when the trending magnitude is no larger than 2, the bias-correction procedure has to be applied to adjust for the bias caused by the problem of endogeneity. When the trending magnitude is greater than two, it seems that the strong trends always guarantee good OLS estimates of the coefficients provided that the sample size is sufficiently large. For instance, Table 4.1 shows that the RMSE is only 0.0242 when the sample size is 1600 in Example 3.

However, statistical inferences of the coefficients under the OLS framework are likely to be severely distorted by the endogeneity issue even for the scenarios with large sample sizes and strong trends ($d > 2$) in the regressors. Namely, the problem of endogeneity renders the reliability of hypothesis tests and confidence intervals even though the OLS estimates are very close to their true values.

We now demonstrate how the issue of endogeneity would affect the performance of the tests. We consider testing the hypothesis

$$\mathbb{H}_0 : \beta = 0.5 \quad \text{vs.} \quad \mathbb{H}_1 : \beta \neq 0.5,$$

where the null hypothesis is true as the value of β is set exactly as 0.5 in our simulation experiment. According to the asymptotic theories, we establish the t-statistics for the hypothesis

$$t = \frac{\widehat{\beta} - 0.5}{s.e.(\widehat{\beta})},$$

where $\widehat{\beta}$ is either the OLS estimator or the bias-corrected estimator, and $s.e.(\widehat{\beta})$ is the corresponding standard error, which could be obtained by the square root of the consistent estimators of the asymptotic variance in Theorems 4.3.1 and 4.32.

Within each replication of the simulation experiment, we compute the t-statistic for the OLS estimator and the bias-corrected estimator. Therefore, under a chosen sample size and trending pattern, we obtain $t_{\text{ols}}^{(1)}, t_{\text{ols}}^{(2)}, \ldots, t_{\text{ols}}^{(N_B)}$ and $t_{\text{bc}}^{(1)}, t_{\text{bc}}^{(2)}, \ldots, t_{\text{bc}}^{(N_B)}$ for the OLS and bias-corrected estimators respectively from N_B independently simulated samples.

When the trending magnitude is smaller than 2 as in Example 1, the endogenous correlation between e_t and v_t causes the t-statistic of the OLS estimator to diverge to infinity. Therefore, we only consider trending patterns in Examples 2 and 3. Based on $N_B = 5000$ replications of the data generating process, the estimated kernel density of the t-statistics for the OLS estimator and the bias-corrected estimator are plotted in Figures 4.2 and 4.3, where the sample sizes are set as 200, 400, 800, and 1600, respectively.

Figure 4.2 shows that when the order of the trending magnitude is 2, the distributions of the t-statistics for the OLS estimators are not centered around zero. Instead, they are always biased and even drift away to the right when the sample size increases. While for the bias-corrected estimators, the distributions of the t-statistics are close to a standard normal distribution whose means are nicely centered around zero. As shown later, bias correction is critical for hypothesis testing as it significantly reduces the probability of making the *type-one error.*

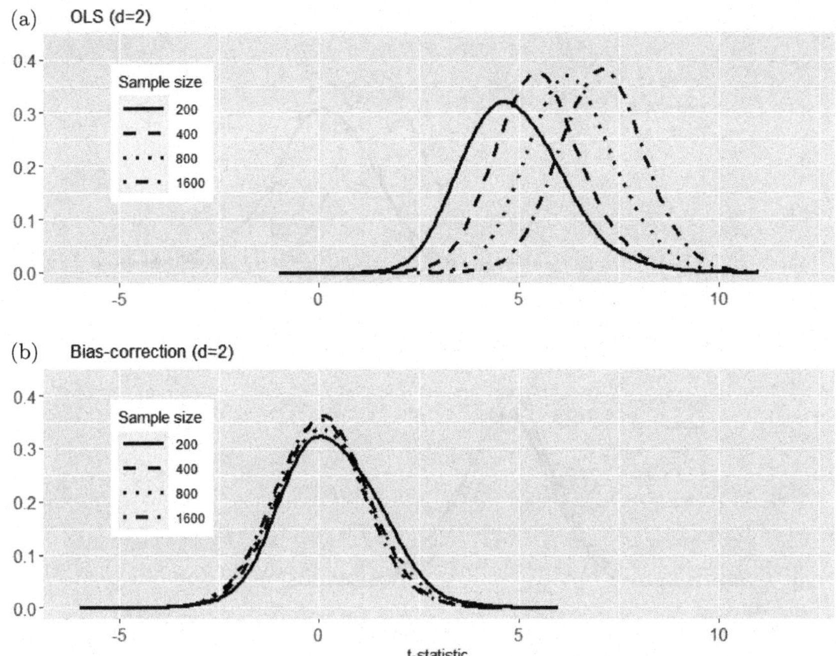

Figure 4.2. The distributions of the t-statistics when $d = 2$.

Figure 4.3 shows the density of the t-statistics for Example 3, where the order of the trending magnitude is 3. Therefore, the OLS estimator is consistent when $T \to \infty$ and the potential bias caused by endogeneity is proportional to $o_P(1)$, which is negligible to the limit distribution. Consequently, in Figure 4.3, all the centers of the t-statistic distributions are moving toward zero for the OLS estimators. However, when the sample size is small, the mean of the t-statistics for the OLS estimator starts from a distant location from zero. Thus, the bias in the distribution of the t-statistic is not negligible. On the other hand, in terms of unbiasedness and consistency, the bias-correction method gives a better estimator where the t-statistics are distributed around zero even for a small sample size.

The adjustments for the endogenous bias are critical to the inference of the coefficients. Note that we should not reject $\mathbb{H}_0$ because the real value of β is 0.5 in the simulation experiment. Table 4.2 shows the proportion of the t-statistic that is greater than the critical value

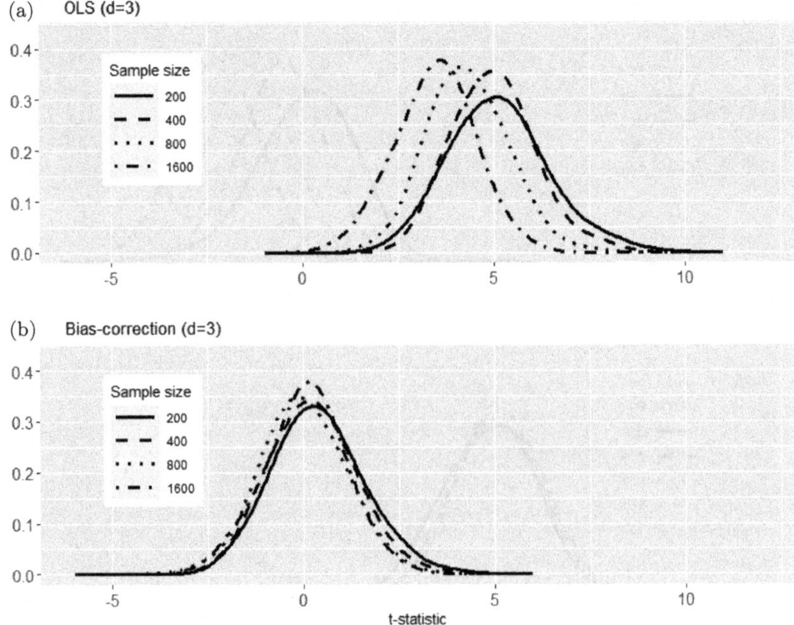

Figure 4.3. The distributions of the t-statistics when $d = 3$.

Table 4.2. The probability of making the *Type I error*.

	Example 2 ($d = 2$)		Example 3 ($d = 3$)	
T	OLS	Bias correction	OLS	Bias correction
200	0.999	0.0933	1.0000	0.0767
400	1.000	0.0760	0.9993	0.0687
800	1.000	0.0680	0.9967	0.0587
1600	1.000	0.0573	0.9420	0.0520

under 5% significance level. Due to the endogenous correlation, the sizes of the t-test based on the simple OLS estimator exhibit severe distortions in both examples. While with a relatively larger sample size, the probability of making the *Type I error* converges to the normal 0.05 for the bias-corrected estimators. To summarize, even though the OLS estimator is consistent when the trending parameter $d > 3$, the inferences are not reliable when endogeneity is present in the regression model. The bias-corrected estimator performs much better in terms of estimations and inferences of the coefficients.

4.6 Empirical Example

In the literature, the logarithm of aggregate income is usually believed to be a random walk process with drift, i.e., $i_t = \omega + i_{t-1} + \nu_{1t}$, for some constant $\omega > 0$. The positive drift generates an upward linear trend, and a pure random walk remains as the residual in the equation $i_t = i_0 + \omega t + z_t$, where $z_t = \sum_{s=1}^{t} \nu_{1s}$. The order of magnitude is $d = 2$ for the pure random walk process that $\sum_{t=1}^{T} z_t^2 = O_P(T^2)$. As discussed in the previous sections, it is difficult to distinguish this realized random walk process from a nonlinear trend-stationary process. Therefore, it is also reasonable to assume that in addition to the linear upward trend, there still exists a time trend that is weaker than the linear time trend but stronger than a weak trend (therefore $1 < d < 3$). Specifically, the data-generating process of log aggregate personal income can be written as $i_t = i_0 + \omega t + g_t + v_t$, where v_t is stationary and g_t is a trending component with magnitude order $1 < d < 3$. Since the strong linear trend ωt is caused by the average growth rate, which is not our concern, we remove the linear trend and focus on the remainder:

$$I_t = g_t + v_t, \tag{4.40}$$

where g_t is a strong trend. For estimation purposes, we further assume that this strong trend can be estimated by the nonparametric kernel method. Specifically, we let $g_t = g_T(\tau_t)$ for $\tau_t = t/T$ when T, the length of the time series, is given. To investigate the long-run MPC, we consider the regression model formulated as

$$C_t = \alpha + I_t\beta + e_t, \tag{4.41}$$

where C_t is aggregate personal consumption but we also remove the strong linear trend in it. We also assume that there exists the problem of endogeneity due to simultaneous causality. Namely, the regression error e_t and the stationary component v_t in I_t are correlated. As we have assumed that I_t is a strongly trending time series, the bias-correction method is applicable because the first-stage OLS estimator is consistent.

Table 4.3 shows the coefficients in model (4.41) estimated by the simple OLS method and the bias-correction method, respectively. The standard errors are also reported in the parenthesis. The table shows that the OLS estimate is about 10% higher than the bias-corrected estimate, indicating that if the issue of endogeneity is

Table 4.3. Estimated coefficients in the strong trending regression model.

	OLS	Bias correction
β	0.7539	0.6857
	(0.0465)	(0.0566)

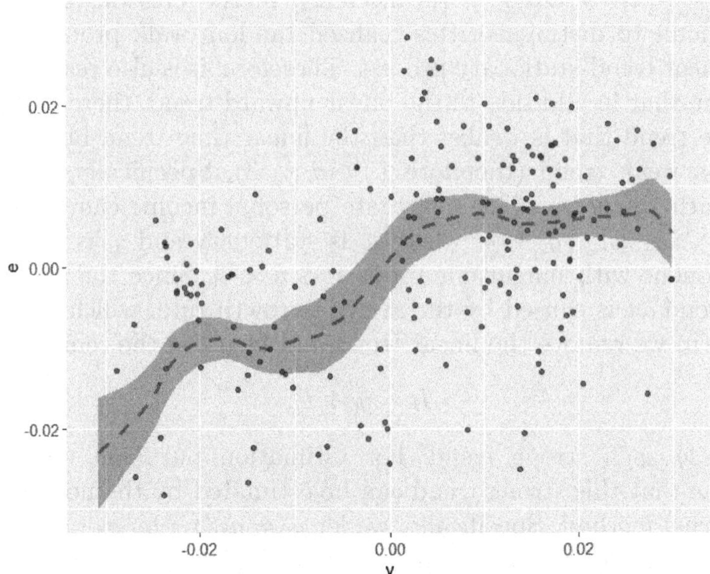

Figure 4.4. Scatter plot of the regression residuals $\widehat{e}_t$ versus $\widehat{v}_t$, and the nonparametric estimation of the endogenous relationship between them.

ignored, we may largely overestimate the long-run MPC. Unit-root tests show that the residual sequence $\widehat{e}_t = C_t - \widehat{\alpha}_{\text{bc}} - I_t\widehat{\beta}_{\text{bc}}$ is stationary with neither stochastic nor deterministic trends.

Figure 4.4 shows the scatter plot of $\widehat{e}_t$ versus the estimated values of $\widehat{v}_t$. They are good approximations of the true values of e_t and v_t if both the trend in I_t and the regression model (4.41) are well estimated. We denote by $\lambda(\cdot)$ a nonparametric function that describes the unknown endogenous relationship between e_t and v_t in the form $e_t = \lambda(v_t) + u_t$. We recover this function nonparametric kernel estimation method and plot their estimated values with 95% confidence bands in Figure 4.4. The positive correlation between e_t and v_t is evident in Figure 4.4. The nonparametric estimates also present a complicated nonlinear manner. To conclude, the OLS estimate of the long-run MPC is positively biased, and the bias-correction procedure proposed in this chapter is needed to eliminate the bias caused by the problem of endogeneity.

4.7 Summary

In this chapter, we investigate the strong trending regression model with endogeneity. Strong trends in the regressors provide dominant information that ensures the consistency of the simple OLS estimators. We show that the asymptotic theories of the OLS estimators depend on the magnitude of trending strength.

Based on the precondition that the OLS estimators are consistent, we propose a bias-correction method that can be applied iteratively to eliminate the endogeneity bias in the OLS estimators. Note that this bias-correction procedure, including parameter estimation and inference, can be applied without the need to know the exact level of trending strength. Numerical simulation results show that the bias-correction method performs quite well in adjusting for the endogeneity bias and significantly reduces the probability of making the type-I error even when the trends are very strong.

4.8 Technical Notes

Proof of Theorem 4.2.1. The estimator $\widehat{Q}$ is defined as $D^{-1}\boldsymbol{X}'\boldsymbol{X}D^{-1}$, then let $d_{ij} = (d_i + d_j)/2 > 1$ and

$$\begin{aligned}
\widehat{Q}_{ij} &= T^{-\frac{d_i+d_j}{2}} \sum_{t=1}^{T} x_{it}x_{jt} = T^{-\frac{d_i+d_j}{2}} \sum_{t=1}^{T} (g_i(t) + v_{it})(g_j(t) + v_{jt}) \\
&= \frac{1}{T^{d_{ij}}} \sum_{t=1}^{T} g_i(t)g_j(t) + \frac{1}{T^{d_{ij}}} \sum_{t=1}^{T} g_i(t)v_{jt} \\
&\quad + \frac{1}{T^{d_{ij}}} \sum_{t=1}^{T} g_j(t)v_{it} + \frac{1}{T^{d_{ij}}} \sum_{t=1}^{T} v_{it}v_{jt}, \qquad (4.42)
\end{aligned}$$

for $i, j = 1, 2, ..., K$. The first term is deterministic and by Assumption 4.2.1, as $T \to \infty$,

$$\frac{1}{T^{d_{ij}}} \sum_{t=1}^{T} g_i(t) g_j(t) \longrightarrow Q_{ij}. \tag{4.43}$$

For the last term,

$$\frac{1}{T^{d_{ij}}} \sum_{t=1}^{T} v_{it} v_{jt} = \frac{1}{T^{d_{ij}-1}} \left(\frac{1}{T} \sum_{t=1}^{T} v_{it} v_{jt} \right), \tag{4.44}$$

where $d_{ij} - 1 > 0$ and $T^{-1} \sum_{t=1}^{T} v_{it} v_{jt} = O_P(1)$. In fact, by Cauchy–Schwarz inequality,

$$\left| \frac{1}{T} \sum_{t=1}^{T} v_{it} v_{jt} \right| \leq \left(\frac{1}{T} \sum_{t=1}^{T} v_{it}^2 \right)^{1/2} \left(\frac{1}{T} \sum_{t=1}^{T} v_{jt}^2 \right)^{1/2}, \tag{4.45}$$

where

$$\begin{aligned} \mathrm{E}\left[\frac{1}{T} \sum_{t=1}^{T} v_{it}^2 \right] &= \mathrm{E}\left[\frac{1}{T} \sum_{t=1}^{T} \left(\sum_{s=0}^{\infty} \psi_{s,i} \eta_{i,t-s} \right)^2 \right] \\ &= \frac{1}{T} \sum_{t=1}^{T} \sum_{s=0}^{\infty} \sum_{l=0}^{\infty} \psi_{s,i} \psi_{l,i} \mathrm{E}\left[\eta_{i,t-s} \eta_{i,t-l} \right] \\ &= \frac{1}{T} \sum_{t=1}^{T} \sum_{s=0}^{\infty} \psi_{s,i}^2 \mathrm{E}\left[\eta_{i,t-s}^2 \right] = \sigma_{ii} \sum_{s=0}^{\infty} \psi_{s,i}^2 < \infty. \end{aligned} \tag{4.46}$$

Therefore, as $T \to \infty$, the last term

$$\frac{1}{T^{d_{ij}}} \sum_{t=1}^{T} v_{it} v_{jt} = O_P(T^{1-d_{ij}}) = o_P(1). \tag{4.47}$$

The second and third terms are similar, so we only study one of them. By Cauchy–Schwarz inequality, it follows that

$$\begin{aligned}\left|\frac{1}{T^{d_{ij}}}\sum_{t=1}^{T} g_i(t)v_{jt}\right| &= \left|\frac{1}{T}\sum_{t=1}^{T}\left(\frac{g_i(t)}{T^{\frac{d_i-1}{2}}}\right)\left(\frac{v_{jt}}{T^{\frac{d_j-1}{2}}}\right)\right| \\ &\leq \left|\frac{1}{T}\sum_{t=1}^{T}\left(\frac{g_i(t)}{T^{\frac{d_i-1}{2}}}\right)^2\right|^{1/2}\left|\frac{1}{T}\sum_{t=1}^{T}\left(\frac{v_{jt}}{T^{\frac{d_j-1}{2}}}\right)^2\right|^{1/2} \\ &= \left|\frac{1}{T^{d_i}}\sum_{t=1}^{T} g_i(t)^2\right|^{1/2}\left|\frac{1}{T^{d_j}}\sum_{t=1}^{T} v_{jt}^2\right|^{1/2}, \end{aligned} \tag{4.48}$$

in which $T^{-d_i}\sum_{t=1}^{T} g_i(t)^2 \longrightarrow Q_{ii} < \infty$ and $T^{-d_j}\sum_{t=1}^{T} v_{jt}^2 = O_P(T^{1-d_j}) = o_P(1)$. Therefore, the second term is $o_P(1)$ and so is the third term. To summarize,

$$\widehat{Q}_{ij} \longrightarrow_P Q_{ij}, \tag{4.49}$$

for all $i, j = 1, 2, \ldots, k$. Therefore, $\widehat{Q} \longrightarrow_P Q$ as $T \to \infty$, in which Q is assumed to be a positive definite matrix.

Proof of Theorem 4.3.1. Recall that the OLS estimator is defined as

$$\widehat{\beta}_{\text{ols}} = \left(\boldsymbol{X}'\boldsymbol{X}\right)^{-1}\left(\boldsymbol{X}'y\right). \tag{4.50}$$

Therefore,

$$D\left(\widehat{\beta}_{\text{ols}} - \beta\right) = \left(D^{-1}\boldsymbol{X}'\boldsymbol{X}D^{-1}\right)^{-1}\left(D^{-1}\boldsymbol{X}'\boldsymbol{e}\right). \tag{4.51}$$

We have already proved in Theorem 4.2.1 that $D^{-1}\boldsymbol{X}'\boldsymbol{X}D^{-1} \longrightarrow_P Q$ as $T \to \infty$, where Q is a positive definite matrix.

Let $\boldsymbol{b} = \mathrm{E}[e_t \boldsymbol{v}_t]$ and $\boldsymbol{b} = (b_1, \ldots, b_k)'$, where for $i = 1, \ldots, k$,

$$b_i = \mathrm{E}[e_t v_{it}] = \mathrm{E}\left[\sum_{s=0}^{\infty} \phi_s \epsilon_{t-s} \sum_{l=0}^{\infty} \psi_{l,i} \eta_{i,t-l}\right]$$
$$= \sum_{s=0}^{\infty} \phi_s \psi_{s,i} \mathrm{E}\left[\epsilon_{t-s} \eta_{i,t-s}\right] = \theta_i \sum_{s=0}^{\infty} \phi_s \psi_{s,i}, \tag{4.52}$$

where $\theta_i = \mathrm{E}[\epsilon_t \eta_{it}]$, which is the ith element of the $k \times 1$ vector $\boldsymbol{\Theta}$. We then focus on $D^{-1}\boldsymbol{X}'\boldsymbol{e}$. Denote

$$D^{-1}\left(\boldsymbol{X}'\boldsymbol{e} - T\boldsymbol{b}\right) \triangleq \boldsymbol{Z}_2. \tag{4.53}$$

Then the ith element of the $k \times 1$ vector $\boldsymbol{Z}_2$ is

$$\begin{aligned}
\boldsymbol{Z}_{2,i} &= T^{-\frac{d_i}{2}} \sum_{t=1}^{T} (x_{it} e_t - b_i) = T^{-\frac{d_i}{2}} \sum_{t=1}^{T} g_i(t) e_t + T^{-\frac{d_i}{2}} \sum_{t=1}^{T} (v_{it} e_t - b_i) \\
&= T^{-\frac{d_i}{2}} \sum_{t=1}^{T} g_i(t) e_t + T^{-\frac{d_i}{2}} \sum_{t=1}^{T} \left(\sum_{s=0}^{\infty} \sum_{l=0}^{\infty} \phi_s \psi_{l,i} \epsilon_{t-s} \eta_{i,t-l} - \sum_{s=0}^{\infty} \phi_s \psi_{s,i} \theta_i \right) \\
&= T^{-\frac{d_i}{2}} \sum_{t=1}^{T} g_i(t) \sum_{s=0}^{\infty} \phi_s \epsilon_{t-s} + T^{-\frac{d_i}{2}} \sum_{t=1}^{T} \sum_{s=0}^{\infty} \phi_s \psi_{s,i} \left(\epsilon_{t-s} \eta_{i,t-s} - \theta_i\right) \\
&\quad + T^{-\frac{d_i}{2}} \sum_{t=1}^{T} \sum_{s=0}^{\infty} \sum_{l=s+1}^{\infty} \phi_s \psi_{l,i} \epsilon_{t-s} \eta_{i,t-l} \\
&\quad + T^{-\frac{d_i}{2}} \sum_{t=1}^{T} \sum_{l=0}^{\infty} \sum_{s=l+1}^{\infty} \phi_s \psi_{l,i} \epsilon_{t-s} \eta_{i,t-l} \\
&\triangleq P_1(T) + P_2(T) + P_3(T) + P_4(T),
\end{aligned} \tag{4.54}$$

for $i = 1, \ldots, K$, where the last three terms correspond to $s = l$, $s < l$, and $s > l$ respectively. Then for $P_1(T)$, as $e_t = \Phi(L)\epsilon_t$, by

Beveridge–Nelson (BN) decomposition, we have

$$
\begin{aligned}
P_1(T) &= T^{-\frac{d_i}{2}} \sum_{t=1}^{T} g_i(t)\Phi(L)\epsilon_t \\
&= T^{-\frac{d_i}{2}} \sum_{t=1}^{T} g_i(t)\Phi(1)\epsilon_t - T^{-\frac{d_i}{2}} \sum_{t=1}^{T} g_i(t)(1-L)\widetilde{\Phi}(L)\epsilon_t \\
&= T^{-\frac{d_i}{2}} \sum_{t=1}^{T} g_i(t)\Phi(1)\epsilon_t + T^{-\frac{d_i}{2}} \sum_{t=1}^{T} g_i(t)\left(\widetilde{\epsilon}_{t-1} - \widetilde{\epsilon}_t\right), \qquad (4.55)
\end{aligned}
$$

where $\widetilde{\epsilon}_t = \widetilde{\Phi}(L)\epsilon_t$. The second term is negligible provided the following lemma holds.

Lemma 4.8.1. *Under Assumptions* 4.2.1–4.2.3, *as* $T \to \infty$,

$$
T^{-\frac{d_i}{2}} \sum_{t=1}^{T} g_i(t)\left(\widetilde{\epsilon}_{t-1} - \widetilde{\epsilon}_t\right) = o_P(1).
$$

For $P_2(T)$, let $f_{i,0}(L) = \sum_{s=0}^{\infty} \phi_s \psi_{s,i} L^s$. By the BN decomposition, we have

$$
\begin{aligned}
P_2(T) &= T^{-\frac{d_i}{2}} \sum_{t=1}^{T} \sum_{s=0}^{\infty} \phi_s \psi_{s,i}\left(\epsilon_{t-s}\eta_{i,t-s} - \theta_i\right) \\
&= T^{-\frac{d_i}{2}} \sum_{t=1}^{T} \sum_{s=0}^{\infty} \phi_s \psi_{s,i} L^s\left(\epsilon_t\eta_{it} - \theta_i\right) = T^{-\frac{d_i}{2}} \sum_{t=1}^{T} f_{i,0}(L)\left(\epsilon_t\eta_{it} - \theta_i\right) \\
&= T^{-\frac{d_i}{2}} \sum_{t=1}^{T} f_{i,0}(1)\left(\epsilon_t\eta_{it} - \theta_i\right) - T^{-\frac{d_i}{2}} \sum_{t=1}^{T} \widetilde{f}_{i,0}(L)(1-L)\left(\epsilon_t\eta_{it} - \theta_i\right) \\
&= T^{-\frac{d_i}{2}} f_{i,0}(1) \sum_{t=1}^{T} \left(\epsilon_t\eta_{it} - \theta_i\right) + T^{-\frac{d_i}{2}} \widetilde{\epsilon\eta}^f_{i,T} - T^{-\frac{d_i}{2}} \widetilde{\epsilon\eta}^f_{i,0}, \qquad (4.56)
\end{aligned}
$$

where we denote $\widetilde{\epsilon\eta}^f_{i,t} = \widetilde{f}_{i,0}(L)(\epsilon_t\eta_{it} - \theta_i)$. The last two terms in (4.56) are negligible according to the following lemma.

Lemma 4.8.2. *Under Assumptions* 4.2.1–4.2.3, *as* $T \to \infty$, *for* $t = 0$ *and* $t = T$,

$$T^{-\frac{d_i}{2}} \widetilde{\epsilon\eta}_{it}^f \longrightarrow_P 0. \tag{4.57}$$

For $P_3(T)$, let $f_{i,q}(L) = \sum_{s=0}^{\infty} \phi_s \psi_{s+q,i} L^s$. By the BN decomposition,

$$\begin{aligned}
&P_3(T)\\
&= T^{-\frac{d_i}{2}} \sum_{t=1}^{T} \sum_{s=0}^{\infty} \sum_{l=s+1}^{\infty} \phi_s \psi_{l,i} \epsilon_{t-s} \eta_{i,t-l}\\
&= T^{-\frac{d_i}{2}} \sum_{t=1}^{T} \sum_{s=0}^{\infty} \sum_{q=1}^{\infty} \phi_s \psi_{s+q,i} \epsilon_{t-s} \eta_{i,t-s-q}\\
&= T^{-\frac{d_i}{2}} \sum_{t=1}^{T} \sum_{q=1}^{\infty} \left(\sum_{s=0}^{\infty} \phi_s \psi_{s+q,i} L^s \right) \epsilon_t \eta_{i,t-q}\\
&= T^{-\frac{d_i}{2}} \sum_{t=1}^{T} \sum_{q=1}^{\infty} f_{i,q}(L) \epsilon_t \eta_{i,t-q}\\
&= T^{-\frac{d_i}{2}} \sum_{t=1}^{T} \sum_{q=1}^{\infty} f_{i,q}(1) \epsilon_t \eta_{i,t-q} - T^{-\frac{d_i}{2}} \sum_{t=1}^{T} \sum_{q=1}^{\infty} \widetilde{f}_{i,q}(L)(1-L) \epsilon_t \eta_{i,t-q}\\
&= T^{-\frac{d_i}{2}} \sum_{t=1}^{T} \sum_{q=1}^{\infty} f_{i,q}(1) \epsilon_t \eta_{i,t-q} - T^{-\frac{d_i}{2}} \sum_{t=1}^{T} (1-L) B_{i,t}^f\\
&= T^{-\frac{d_i}{2}} \sum_{t=1}^{T} \sum_{q=1}^{\infty} f_{i,q}(1) \epsilon_t \eta_{i,t-q} + T^{-\frac{d_i}{2}} B_{i,0}^f - T^{-\frac{d_i}{2}} B_{i,T}^f,
\end{aligned} \tag{4.58}$$

where $B_{i,t}^f = \sum_{q=1}^{\infty} \widetilde{f}_{i,q}(L) \epsilon_t \eta_{i,t-q}$. The last two terms are negligible provided the following lemma holds.

Lemma 4.8.3. *Under Assumptions* 4.2.1–4.2.3, *as* $T \to \infty$, *for* $t = 0$ *and* $t = T$,

$$T^{-\frac{d_i}{2}} B_{i,t}^f \longrightarrow_P 0. \tag{4.59}$$

Similarly, for $P_4(T)$, let $m_{i,q}(L) = \sum_{l=0}^{\infty} \phi_{q+l}\psi_{l,i}L^l$. By the BN decomposition,

$$
\begin{aligned}
P_4(T) &= T^{-\frac{d_i}{2}} \sum_{t=1}^{T} \sum_{l=0}^{\infty} \sum_{s=l+1}^{\infty} \phi_s \psi_{l,i} \epsilon_{t-s} \eta_{i,t-l} \\
&= T^{-\frac{d_i}{2}} \sum_{t=1}^{T} \sum_{l=0}^{\infty} \sum_{q=1}^{\infty} \phi_{q+l} \psi_{l,i} \epsilon_{t-q-l} \eta_{i,t-l} \\
&= T^{-\frac{d_i}{2}} \sum_{t=1}^{T} \sum_{q=1}^{\infty} \left(\sum_{l=0}^{\infty} \phi_{q+l} \psi_{l,i} L^l \right) \epsilon_{t-q} \eta_{i,t} \\
&= T^{-\frac{d_i}{2}} \sum_{t=1}^{T} \sum_{q=1}^{\infty} m_{i,q}(L) \epsilon_{t-q} \eta_{i,t} \\
&= T^{-\frac{d_i}{2}} \sum_{t=1}^{T} \sum_{q=1}^{\infty} m_{i,q}(1) \epsilon_{t-q} \eta_{i,t} - T^{-\frac{d_i}{2}} \sum_{t=1}^{T} \sum_{q=1}^{\infty} \widetilde{m}_{i,q}(L)(1-L) \epsilon_{t-q} \eta_{i,t} \\
&= T^{-\frac{d_i}{2}} \sum_{t=1}^{T} \sum_{q=1}^{\infty} m_{i,q}(1) \epsilon_{t-q} \eta_{i,t} - T^{-\frac{d_i}{2}} \sum_{t=1}^{T} (1-L) B_{i,t}^{m} \\
&= T^{-\frac{d_i}{2}} \sum_{t=1}^{T} \sum_{q=1}^{\infty} m_{i,q}(1) \epsilon_{t-q} \eta_{i,t} + T^{-\frac{d_i}{2}} B_{i,0}^{m} - T^{-\frac{d_i}{2}} B_{i,T}^{m}, \qquad (4.60)
\end{aligned}
$$

where $B_{i,t}^m = \sum_{q=1}^{\infty} \widetilde{m}_{i,q}(L)\epsilon_{t-q}\eta_{i,t}$. According to the following lemma, we can ignore the last two terms in (4.60).

Lemma 4.8.4. *Under Assumptions* 4.2.1–4.2.3, *as* $T \to \infty$, *for* $t = 0$ *and* $t = T$,

$$T^{-\frac{d_i}{2}} B_{i,t}^{m} \longrightarrow_P 0. \qquad (4.61)$$

To summarize, we can ignore the negligible terms in $P_i(T)$ for $i = 1, 2, 3, 4$. Thus, equation (4.54) can be written as

$$\boldsymbol{Z}_{2,i} = \sum_{t=1}^{T} M_{Tt}^{i} + o_P(1), \qquad (4.62)$$

where

$$M_{Tt}^{i}=T^{-\frac{d_i}{2}}\left(g_i(t)\Phi(1)\epsilon_t+\sum_{q=1}^{\infty}f_{i,q}(1)\epsilon_t\eta_{i,t-q}+f_{i,0}(1)\left(\epsilon_t\eta_{it}-\theta_i\right)\right.$$
$$\left.+\sum_{q=1}^{\infty}m_{i,q}(1)\epsilon_{t-q}\eta_{i,t}\right). \tag{4.63}$$

We denote $M_{Tt}=(M_{Tt}^{1},\ldots,M_{Tt}^{K})'$, then

$$\boldsymbol{Z}_2=\sum_{t=1}^{T}M_{Tt}+o_P(1). \tag{4.64}$$

M_{Tt} is a martingale difference sequence (m.d.s.) suggested by the equation as follows:

$$\mathrm{E}\left[M_{Tt}|F_{t-1}\right]=0, \tag{4.65}$$

where F_{t-1} is the filtration that $F_{t-1}=\{\epsilon_{t-1},\epsilon_{t-2},\ldots,\boldsymbol{\eta}_{t-1},\boldsymbol{\eta}_{t-2},\ldots\}$. We are able to apply the central limit theorem for martingale difference sequence (m.d.s) given the following lemma holds.

Lemma 4.8.5. *For any $K\times 1$ vector $a=(a_1,\ldots,a_K)'$ with $||a||=1$,*

$$\sum_{t=1}^{T}\mathrm{E}\left[\left(a'M_{Tt}\right)^2\Big|F_{t-1}\right]\longrightarrow_P a'\Omega a \tag{4.66}$$

and

$$\sum_{t=1}^{T}\mathrm{E}\left[\left(a'M_{Tt}\right)^4\Big|F_{t-1}\right]\longrightarrow_P 0, \tag{4.67}$$

where Ω is a $K\times K$ positive definite matrix $\Omega_{ij}=\sigma_1^2\Phi(1)^2Q_{ij}$.

Remark 4.8.1. Once the two conditions are satisfied, we can apply the CLT for $a'M_{Tt}$, which is any linear combination of the elements in M_{Tt}. Thus, the vector $\sum_{t=1}^{T}M_{Tt}$ converges in distribution to multivariate Gaussian:

$$\sum_{t=1}^{T}M_{Tt}\longrightarrow_D\mathcal{N}(0,\Omega). \tag{4.68}$$

Therefore, by equation (4.64), we have

$$\boldsymbol{Z}_2 = D^{-1}(\boldsymbol{X}'\boldsymbol{e} - T\boldsymbol{b}) \longrightarrow_D \mathcal{N}(0, \Omega), \tag{4.69}$$

with $\Omega_{ij} = \sigma_1^2 \Phi(1)^2 Q_{ij}$. As $T \to \infty$, since $\widehat{Q} = D^{-1}\boldsymbol{X}'\boldsymbol{X}D^{-1} \longrightarrow_P Q$, we have

$$\begin{aligned} D\left(\widehat{\beta}_{\text{ols}} - \beta\right) - \widehat{Q}^{-1}D^{-1}T\boldsymbol{b} &= (D^{-1}\boldsymbol{X}'\boldsymbol{X}D^{-1})^{-1}(D^{-1}(\boldsymbol{X}'e - T\boldsymbol{b})) \\ &\longrightarrow_D \mathcal{N}(0, Q^{-1}\Omega Q^{-1}), \end{aligned} \tag{4.70}$$

i.e.,

$$D\left(\widehat{\beta}_{\text{ols}} - \beta - D^{-1}\widehat{Q}^{-1}D^{-1}T\boldsymbol{b}\right) \longrightarrow_D \mathcal{N}(0, Q^{-1}\Omega Q^{-1}). \tag{4.71}$$

Thus, we complete the proof of this theorem. ■

Chapter 5

Testing for Common Trends

> ... cointegration regressions do not explain trends. Instead, they relate trends in multiple time series ... trends themselves can be validly modeled in a variety of ways.
>
> — Phillips (1998)

Economic and financial theories occasionally predict the existence of shared common trends among specific time series variables. This implies that a linear or nonlinear combination of nonstationary trending time series can result in a stationary series. This phenomenon suggests that these nonstationary variables move together in the long run, establishing a stable equilibrium relationship. It's important to note that these common trend relationships are long-run associations, differing from correlations between short-run shocks or innovations. The key distinction lies in the fact that economic forces responding to deviations from equilibrium may require a relatively extended period to restore equilibrium.

There are many empirical examples of common trend relations based on economic, financial, or other theories. A few examples are as follows. The purchasing power parity (PPP) implies that the nominal exchange rate and foreign and domestic prices share a common trend according to *the law of one price* in the absence of trade frictions and under conditions of free competition and price flexibility. Similarly, covered interest rate parity (CIRP) implies a common trend between forward and spot exchange rates when the no-arbitrage condition is satisfied. The expectations hypothesis of the term structure implies that nominal interest rates at different maturities may share a common trend. In climate change, if global warming is caused by excess emissions of greenhouse gases from anthropogenic activities, then the time series of global temperatures and greenhouse gas emissions would share a common trend, indicating a long-run stable relationship between them.

In this chapter, we first introduce the concept of co-trending and then establish a unified framework – a coordinate system – to capture and represent trends in nonstationary time series. Further, a common features approach is developed to test for common trends among trending time series, in which the nature of trends is unknown. Finally, simulated examples are provided to show the performance of the test.

5.1 Testing for a Common Deterministic Trend

In many practical applications, deterministic trends are usually restricted to linear specifications. For example, to examine economic convergence, Vogelsang and Franses (2005) considered a simple model with n trend-stationary time series,

$$y_{it} = \alpha_i + \beta_i t + e_{it}, \tag{5.1}$$

for $i = 1, 2, \ldots, n$, $t = 1, 2, \ldots, T$, where each e_{it} is a stationary I(0) process, but they are allowed to be contemporaneously correlated with the covariance matrix Ω. We are interested in testing whether β_i's are the same for all or some of the time series so that they exhibit deterministic linear time trends with the same slope. According to Grenander and Rosenblatt (2008), the OLS estimator of β, for $\beta = (\beta_1, \ldots, \beta_n)$, is asymptotically equivalent to the GLS estimator when e_i's are second-order stationary. Since equation (5.1) is a seemingly unrelated regression (SUR) with the same regressor t in each equation, the OLS estimator is equivalent to the SUR estimator, which is the GLS estimator that accounts for the contemporaneous correlation between the series. Specifically, as $T \to \infty$, the OLS estimator

$$T^{3/2}(\widehat{\beta} - \beta) \longrightarrow_D \mathcal{N}(0, 12\Omega), \tag{5.2}$$

which converges at the rate of $O(T^{3/2})$. However, reliable inference of β highly depends on the consistent estimation of Ω, which has been extensively studied in the literature. For example, after obtaining the residual vector $\widehat{e}_t = (\widehat{e}_{1t}, \ldots, \widehat{e}_{nt})'$ using $\widehat{e}_{it} = y_{it} - \widehat{\alpha}_i - \widehat{\beta}_i t$, the HAC estimators following Newey and West (1987) and Andrews (1991) using the Bartlett kernel are established as $\widehat{\Omega} = \widehat{\Gamma}_0 + \sum_{j=1}^{p}(1-j/p)(\widehat{\Gamma}_j + \widehat{\Gamma}_j')$, where $\widehat{\Gamma}_j = T^{-1}\sum_{t=j+1}^{T} \widehat{e}_t \widehat{e}_{t-j}$ with p being a truncation parameter satisfying $p \to \infty$ and $p/T \to 0$ as $T \to \infty$. Sun (2011) developed a novel testing procedure for hypotheses on deterministic trends in a multivariate trend-stationary model, where the long-run variance is estimated using the series method with carefully selected basis functions. In particular, $\widehat{\Omega} = p^{-1}\sum_{j=1}^{p} \widehat{\Omega}_j$, where $\widehat{\Omega}_j = \widehat{\Gamma}_j \widehat{\Gamma}_j'$ and $\widehat{\Gamma}_j = T^{-1/2}\sum_{t=1}^{T} \phi_j(\tau_t)\widehat{e}_t$, in which $\{\phi_j(r)\}$ $(r \in [0,1])$ is an orthonormal sequence of eigenfunctions, such as $\{\phi_j(r) = \sqrt{2}\cos \pi j r, j = 0, 1, \ldots\}$.

The two papers proposed a general test for $\mathbb{H}_0 : R\beta = r$ against $\mathbb{H}_1 : R\beta \neq r$, and R and r are $q \times n$ and $q \times 1$ matrices of known constants, respectively, where q is the number of joint restrictions. The test contains the special case of common deterministic trend slopes. For example, $\mathbb{H}_0 : \beta_1 = \cdots = \beta_m = \beta$, for some β and $1 \leq m \leq n$. In other words, we are able to use this test to examine whether linear trends in all or some of the trend-stationary time series have the same slope. Within the same framework, Xu (2012) considered similar tests for multivariate deterministic trend coefficients in the case of a nonstationary variance process.

In a broader sense, linear combinations of deterministically trending time series may be covariance stationary so that their trends cancel out in the long run. Bierens (2000) and Düker *et al.* (2022) defined this as a *co-trending* relationship within a vector of k deterministically trending time series $y_t = (y_{1t}, \ldots, y_{kt})'$. Specifically, they assume that

$$y_t = \mu(\tau_t) + e_t, \tag{5.3}$$

for $t = 1, 2, \ldots, T$, where $\mu(\tau_t) = (\mu_1(\tau_t), \ldots, \mu_k(\tau_t))'$ represents their time trends or varying means and $e_t = (e_{1t}, \ldots, e_{kt})'$ are error terms with zero mean and finite variance. The co-trending relationship within y_t indicates the existence of a $k \times d$ matrix B with linearly independent columns such that

$$B'\mu(u) = \mu, \tag{5.4}$$

for $u \in (0, 1]$, and μ is a $d \times 1$ constant vector after the combinations. Düker *et al.* (2022) also defined that the largest value of d for which (5.4) holds is called the *co-trending dimension* of the corresponding co-trending subspace $\mathcal{B}$, spanned by the columns of matrix B. Meanwhile, $d_2 = k - d$ is called the *non-cotrending dimension* of the corresponding non-cotrending subspace $\mathcal{B}_2$, with $\mathcal{B}_2 \perp \mathcal{B}$. Note that the dimension d_2 indicates how many non-constant deterministic functions drive the system of (5.4). In this context, time series with linear time trends can be regarded as co-trending because they share a common trend automatically, except for different trend slopes. By demeaning the time trends or varying means, we can establish an integral $M = \int_0^1 (\mu(u) - \bar{\mu})(\mu(u) - \bar{\mu})' du$ with $\bar{\mu} = \int_0^1 \mu(u) du$. Then, the co-trending relationship in equation (5.4) can be transformed into

$$B'MB = 0, \tag{5.5}$$

and $d_2 = \text{rank}(M)$ and $d = k - d_2$. This shows that d and d_2 are the nullity and rank of the matrix M, and the co-trending subspace $\mathcal{B}$ is spanned by the eigenvectors associated with the zero eigenvalues of M. The remaining problem is to test for the rank of matrix M, which can be approximated by a consistent estimator. Düker *et al.* (2022) employed a symmetric estimator: $\widehat{M}_s = (\widehat{M} + \widehat{M}')/2$, where $\widehat{M} = T^{-1} \sum_{t=1}^{T-1} (y_t - \bar{y})(y_{t+1} - \bar{y})'$ with $\bar{y} = \sum_{t=1}^{T} y_t / T$.

5.2 Coordinate Systems for Trends

In practical scenarios, the exact nature of trends in nonstationary time series data is often unknown, especially when working with a limited sample size. This challenge becomes pronounced when attempting to establish a unified framework capable of capturing various types of trends, including stochastic trends, deterministic trends, and their combinations, behind the trending time series we observe in the empirical problems. The key question then arises: Can we establish a comprehensive approach that accommodates diverse embedded trends that are not precisely distinguished?

In explaining spurious regression, Phillips (1998) provided some new insights into representing different types of trends. He established a coordinate system $\{\phi_k\}_{k=1}^{\infty}$, which is a complete orthogonal system in $L_2[0,1]$. They are basis functions of τ_t, for $t = 1, 2, \ldots, T$, satisfying $\int_0^1 \phi_k(\tau)\phi_l(\tau)d\tau = \delta_{kl}$, where δ_{kl} is the Kronecker delta defined as $\delta_{kl} = 1$ if $k = l$ and $\delta_{kl} = 0$ if $k \neq l$.

First, such deterministic basis functions of time can be used to represent stochastic trends. Specifically, we consider the regression equation that projects a random trend onto the deterministic basis functions as

$$\frac{y_t}{\sqrt{T}} = \sum_{k=1}^{K} \widehat{a}_k \phi_k \left(\frac{t}{T}\right) + \frac{\widehat{u}_t}{\sqrt{T}}, \tag{5.6}$$

for $t = 1, 2, \ldots, T$, where $y_t = \sum_{s=1}^{t} \varepsilon_s$ is a random walk process in which ε_t is a stationary time series with zero mean and finite absolute moments to order $p > 2$. The regressors are the first K basis functions, and $\widehat{a}_1, \widehat{a}_2, \ldots, \widehat{a}_K$ are fitted coefficients estimated by the least squares method. The limiting behavior of the dependent variable is a Brownian motion given by

$$\frac{y_{[Tr]}}{\sqrt{T}} = \frac{1}{\sqrt{T}} \sum_{s=1}^{[Tr]} \varepsilon_s \Rightarrow B(r). \tag{5.7}$$

There are several major findings regarding this regression relationship, as emphasized by Phillips (1998):

(1) The fitted coefficients $\widehat{a} = (\widehat{a}_1, \ldots, \widehat{a}_K)$ turned out to be random variables instead of converging to certain constants. In other words, stochastic trends are represented by nonrandom regressors with random coefficients. Specifically, $c_K'\widehat{a}$ converges to a normal distribution where $c_K' \in \mathbb{R}^K$ is any vector with $c_K' c_K = 1$ if K is fixed or $K \to \infty$ and $K/T \to 0$ as $T \to \infty$.

(2) The usual t-ratios of the estimated coefficients diverge at the rate of $O(\sqrt{T})$ so that all the coefficients would be significant for sufficiently large T. This shows that the stochastic trend in y_t is captured by the basis functions on $[0,1]$, with significant projections on a certain "axis" of the coordinate system.

(3) When $K \to \infty$ and $K/T \to 0$ as $T \to \infty$, R^2 of the regression converges to 1, indicating that y_t could be accurately reproduced by sufficiently many deterministic basis functions. The above results hold as well when y_t is a general continuous stochastic process:

$$\frac{y_t}{T^\alpha} = \sum_{k=1}^{K} \widehat{a}_k \phi_k \left(\frac{t}{T}\right) + \frac{\widehat{u}_t}{T^\alpha}, \tag{5.8}$$

for some $\alpha > 0$.

Second, continuous deterministic trends could also be approximated by a linear combination of Wiener processes. Specifically, for a continuous and square-integrable function $f(\cdot)$ on $[0,1]$, we can approximate it by independent standard Brownian motions $\{W_i\}_{i=1}^N$, such that as $N \to \infty$,

$$\sup_{r\in[0,1]} \left| f(r) - \sum_{i=1}^{N} d_i W_i(r) \right| < \varepsilon, \tag{5.9}$$

almost surely, where ε is an arbitrarily small positive real value, $d_1, \ldots, d_N$ are random coefficients. Therefore, we have

$$f(r) \sim \sum_{i=1}^{\infty} d_i W_i(r) \tag{5.10}$$

in L_2. To establish a coordinate system by making the series of $\{W_i(r)\}_{i=1}^N$ orthogonal, one can employ the Gram–Schmidt process to orthogonalize $\{W_i(r)\}_{i=1}^N$ to $\{V_i(r)\}_{i=1}^N$. Specifically,

$$\begin{cases} V_1 = W_1, \\ V_2 = W_2 - \tilde{W}_2, \\ V_3 = W_3 - \tilde{W}_3, \\ \vdots \end{cases} \tag{5.11}$$

where $\tilde{W}_i$ is the projection of W_i on $(V_1, \ldots, V_{i-1})'$ for $i = 2, 3, \ldots, N$. For example, $\tilde{W}_2 = V_1(\int_0^1 V_1^2(r)dr)^{-1}(\int_0^1 V_1(r)W_2(r)dr)$, and similarly, we have $\tilde{W}_3 = V_{12}'(\int_0^1 V_{12}(r)V_{12}(r)'dr)^{-1}(\int_0^1 V_{12}(r)W_3(r)dr)$, in which $V_{12} = (V_1, V_2)'$. This means that when T is large, a deterministic trend $f(\tau_t)$ accumulates information and becomes dense on $[0, 1]$. The continuous function of $f(r)$ on $[0, 1]$ could be well approximated by $\{V_i\}_{i=1}^N$, a set of orthonormal stochastic trends. The coefficients are random and can be estimated by the least squares method.

Third, Phillips (1998) also showed that a Brownian motion (i.e., a stochastic trend) on the interval $[0, 1]$ can also be approximated by independent Brownian motions $W_i(r)$, $i = 1, 2, \ldots$, given by

$$B(r) \sim \sum_{i=1}^{\infty} d_i W_i(r), \tag{5.12}$$

in $L_2[0, 1]$ almost surely, where d_i's are random coefficients. While using the orthogonal basis $\{V_i\}_{i=1}^{\infty}$, we have

$$B(r) \sim \sum_{i=1}^{\infty} \tilde{d}_i V_i(r), \tag{5.13}$$

where the coefficients can be conveniently estimated in a separate way using $\tilde{d}_i = (\int_0^1 V_i^2(r)dr)^{-1} \int_0^1 V_i(r)B(r)dr$, for $i = 1, 2, \ldots$. Equation (5.13) conveys the message that stochastic trends can be represented by a stochastic basis in addition to equation (5.8) using a deterministic basis.

Fourth, according to the Weierstrass approximation theorem, any continuous function $f(r)$ can be uniformly approximated on the interval $[0, 1]$ by the basis functions $\phi_k(r)$. For example, using a sequence of trigonometric polynomials, we have

$$f(r) \sim \alpha_0 + \sum_{k=1}^{K} (\alpha_k \sin(2\pi k r) + \beta_k \cos(2\pi k r)), \tag{5.14}$$

where α_k and β_k are nonrandom coefficients. In fact, everything in equation (5.14) is fixed, and this is the case that deterministic trends can be approximated by deterministic basis functions.

To summarize, we can always establish a trend coordinate system consisting of either stochastic or deterministic orthogonal basis functions. By projecting the empirical time series onto such a coordinate system, the embedded trends of any nature can be well captured and represented. Namely, trends in the empirical time series could be represented in a variety of ways. This theory somehow explains why spurious regression may occur when significant associations are found between two or more independent unit-root series. The reason is that the stochastic trend in one integrated time series is captured and represented by trends in other integrated time series. In fact, according to the theory by Phillips (1998), spurious regression may occur among any trending time series, no matter whether their trends are deterministic or stochastic. Meanwhile, this finding also helps in understanding some of the size or power issues in unit-root tests in which we may mistakenly reject the null in favor of the alternative, while, in fact, the alternative is nothing but an alternative representation of the unit-root null.

5.3 A Common Features Approach

We acknowledge the difficulty in determining the exact nature of trends in nonstationary trending time series. Moreover, we consider the possibility of endogeneity in the co-trending regression equation and propose a method to address this issue. The problem of endogeneity can arise when the regressand and regressors in a regression model are simultaneously determined. This can affect the quality of the least squares estimator in the time series regression. Despite the super-consistency of the OLS estimator of cointegrating vectors, endogeneity can cause significant finite sample biases that can lead to incorrect inferences. The extent of this finite sample bias caused by endogeneity has been studied by Phillips and Hansen (1990) for cointegrating regressions, Elliott (1998) for regressions involving time series with near unit root, and Chen (2017) for regressions involving time series with general trends. We approach both the issue of ambiguity over the nature of the trend and the problem of endogeneity from a common features perspective (Engle and Kozicki, 1993). A feature is a dominant statistical property that satisfies the following axioms:

- if x_t has (does not have) the feature, then cx_t, where $c \neq 0$ will also have (not have) it;
- if x_t and y_t do not have the feature, $x_t + y_t$ will not have it;
- if x_t has the feature and y_t does not have it, then $x_t + y_t$ will have the feature.

For example, a trend, whether deterministic, stochastic, or a mixture of deterministic and stochastic, is a feature. Also, serial correlation (Engle and Kozicki, 1993), business cycles (Vahid and Engle, 1993), seasonality (Engle and Hylleberg, 1996), and nonlinearity (Anderson and Vahid, 1998) can be regarded as features. Two series have a common feature if both of them have the same feature, but a linear combination of them does not have that feature. In our context, the feature of interest is a general trend without restricting it to being of a specific nature. When y_t and x_t have a general trend, our objective is to establish if their trend is common and, if it is, to provide an estimator for β_0 in their co-trending relationship, $y_t - \beta_0 x_t$, the linear combination that does not have a trend. These time series might have both deterministic and stochastic trends; however, our definition requires that there be a linear combination that eliminates both.

Suppose that y_t and x_t have the same feature F. Under the null that F is a common feature between y_t and x_t, there is one and only one β such that $e_t = y_t - x_t\beta$ does not have feature F. Specifically, F is taken as a "trend," and we consider trending time series of the form

$$y_t = f_t + u_t, \tag{5.15}$$

$$x_t = g_t + v_t, \tag{5.16}$$

for $t = 1, 2, \ldots, T$, where f_t and g_t are trending components, and u_t and v_t are ergodic stationary disturbances. As it is difficult to determine the exact nature of trends in nonstationary trending time series x_t and y_t, f_t and g_t are not specified to take certain forms, nor are they estimable from observed values of y_t and x_t. Therefore, the trend can be linear or a finite-order polynomial of time; it can be an integrated stochastic process of order 1 or larger; it can be a piecewise linear trend; or it can even be a combination of these forms. It is important to be clear that the possibility that these series may have a common stochastic trend but different deterministic trends is of no interest to us. If the trends in y_t and x_t are common, then there exists one and only one β_0 such that $f_t = \beta_0 g_t$, and there is no trend in $y_t - \beta_0' x_t$. Our objective is to estimate β_0 and derive its asymptotic distribution.

To this end, we consider the pseudo-structural model

$$y_t = \alpha_0 + \beta_0' x_t + e_t, \tag{5.17}$$

$$x_t = g_t + v_t, \tag{5.18}$$

where g_t is a trend process. We call this a pseudo-structural system because e_t and v_t are correlated, and hence the right-hand-side variable in the first equation is endogenous. We first focus on the consistent estimation of β_0 in this pseudo-structural system.

The first argument for our proposed estimation method is that the powers of $\tau_t = (t/T)$ are legitimate instruments that can produce an instrumental variable (IV) estimator of β_0 in equation (5.17), regardless of whether the trend (g_t) is deterministic or stochastic. That powers of τ_t can be used as instruments in estimating cointegrating relationships is evident in Park and Phillips (1988), and the performance of this IV estimator in the case of I(1) variables has been studied in Phillips and Hansen (1990). Here, we propose that this applies more generally to all forms of common trends. The rationale is based on Phillips (1998).

A corollary of Theorem 3.1 in Phillips (1998) is that if y_t/n^α has the same limiting behavior as a continuous-time stochastic process on $[0, 1]$ for some $\alpha > 0$, then the regression

$$y_t = \sum_{k=1}^{K} \widehat{b}_k \phi_k(\tau_t) + \widehat{u}_t \tag{5.19}$$

correctly signals that these basis functions are statistically significant. This occurs because a continuous stochastic process on $[0, 1]$ has a legitimate orthogonal representation in a coordinate system given by deterministic functions of time.

Note that equation (5.19) is merely a means for obtaining a consistent estimate of β_0 in equation (5.17). It does not suggest that the true trend in y_t is a deterministic polynomial function of time or that equation (5.19) should be used for forecasting y_t. The issues involved in forecasting with trending variables are discussed in Chapter 20 of Elliott and Timmermann (2016). Therefore, such polynomial functions of time are relevant instruments. Such deterministic functions of time are not significant predictors of non-trending variables, and hence they satisfy both requirements of legitimate instruments.

The instruments could be selected as the first $(p+1)$ elements from a set of orthogonal basis functions. For example, one can use the normalized *Legendre polynomials* defined on $[0,1]$, which take the form

$$\phi_j(\tau) = \sqrt{2j+1}\sum_{s=0}^{j}(-1)^{j+s}\binom{j}{s}\binom{j+s}{s}\tau^s \quad \text{for } j = 0,1,2,\ldots. \tag{5.20}$$

The IVs satisfy $\int_0^1 \phi_i(\tau)\phi_j(\tau)d\tau = \delta_{ij}$, where $\delta_{ij} = 1$ for $i = j$ and $\delta_{ij} = 0$ for $i \neq j$, $i, j = 0, 1, \ldots, p$. With $\tau_t = t/T$, these will have a significant relationship (i.e., a *spurious correlation*) with the trending components in x_t regardless of whether the trend is deterministic or stochastic.

We could also adopt other orthogonal polynomials, such as the *Chebyshev polynomials.* As the column spaces of the first $p+1$ elements of all such polynomials are the same as the column spaces of a matrix whose row t is $\{1, t, t^2, \ldots, t^p\}$ for all values of p, the choice of orthogonal basis polynomials does not affect the two-stage least squares estimator of β_0. In fact, the use of an orthogonal basis function is for theoretical considerations, and it also speeds up simulations.

In order to achieve exact orthogonality in finite samples, one can use the normalized *discrete Legendre polynomials* (Neuman and Schonbach, 1974) given by[1]

$$\psi_j(t,T) = \frac{\sqrt{T}}{M_j}\sum_{s=0}^{j}(-1)^s\binom{j}{s}\binom{j+s}{s}\frac{t(t-1)\cdots(t-s+1)}{T(T-1)\cdots(T-s+1)}, \tag{5.21}$$

where $M_j = \left(\frac{1}{2j+1}\frac{(T+j+1)(T+j)\cdots(T+1)}{T(T-1)\cdots(T-j+1)}\right)^{1/2}$ for $t = 0, 1, \ldots, T$ and $j = 0, 1, 2, \ldots$. The discrete basis functions $\psi_j(t,T)$ satisfy $T^{-1}\sum_{t=0}^{T}\psi_i(t,T)\psi_j(t,T) = \delta_{ij}$, where δ_{ij} was defined above. Figure 5.1 shows the first five *Legendre polynomials.*

[1]The time index spans from 0 to n and forms $n+1$ equally spaced points.

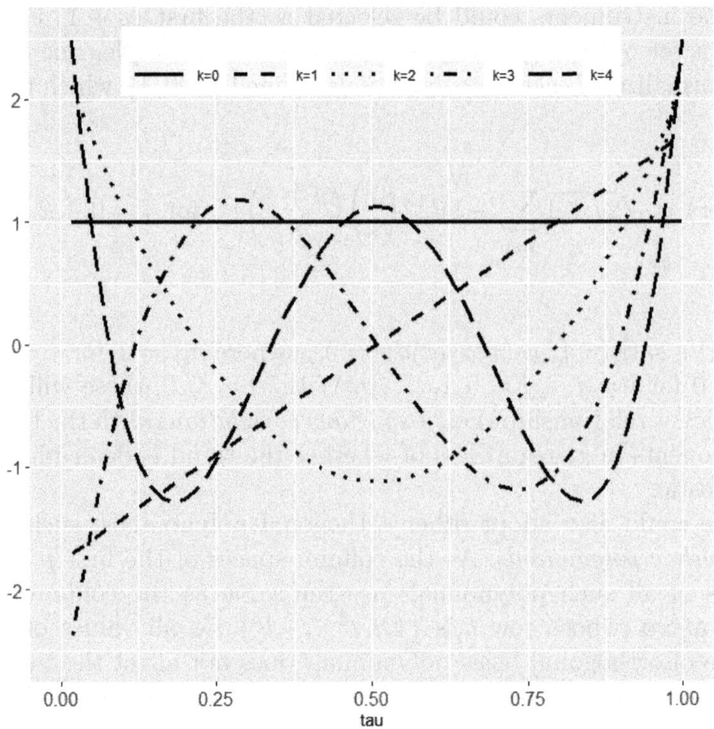

Figure 5.1. The plot of first five *Legendre polynomials* on $[0,1]$.

Meanwhile, under the null hypothesis of common trend, $e_t = y_t - \alpha_0 - x_t'\beta_0$ is merely a zero-mean stationary process without any trend. Therefore, e_t is "*uncorrelated*" with the basis functions so that $(\phi_0(\tau), \phi_1(\tau), \ldots, \phi_p(\tau))$ can be used as instruments to estimate (α_0, β_0') based on the asymptotic orthogonality conditions

$$\frac{1}{T}\sum_{t=1}^{T} \phi_j(\tau_t) e_t \to_P 0, \quad \text{as } T \to \infty, \tag{5.22}$$

for $j = 0, 1, 2, \ldots, p$, where $e_t = y_t - \alpha_0 - x_t'\beta_0$. Theoretically, p can be any finite integer larger than the dimension of β_0 (i.e., β_0 needs to be overidentified). However, in practice, a very large p can incidentally fit the stationary variations of e_t in a small sample. Hence, when the sample size is small, we suggest choosing the smallest p that is larger than the dimension of β_0.

In matrix notation, let $y = (y_1, y_2, \ldots, y_T)'$, $e = (e_1, e_2, \ldots, e_T)'$, $\gamma_0 = (\alpha_0, \beta_0')'$,

$$X = \begin{pmatrix} 1 & x_{11} & x_{21} & \cdots & x_{k1} \\ \vdots & \vdots & \vdots & \ddots & \vdots \\ 1 & x_{1T} & x_{2T} & \cdots & x_{kT} \end{pmatrix}, \tag{5.23}$$

and

$$\Phi = \begin{pmatrix} 1 & \phi_1(\tau_1) & \phi_2(\tau_1) & \cdots & \phi_p(\tau_1) \\ \vdots & \vdots & \vdots & \ddots & \vdots \\ 1 & \phi_1(\tau_n) & \phi_2(\tau_T) & \cdots & \phi_p(\tau_T) \end{pmatrix}, \tag{5.24}$$

where $\tau_t = t/T$ for $1 \le t \le T$. Then, the regression equation can be written as

$$y = X\gamma_0 + e. \tag{5.25}$$

When orthogonal basis functions in the columns in Φ are taken as IVs, the *two-stage least squares estimator* of the parameters is

$$\widehat{\gamma}_{2sls} = (\widehat{\alpha}_{2sls}, \widehat{\beta}_{2sls}')' = (X'\Phi(\Phi'\Phi)^{-1}\Phi'X)^{-1}X'\Phi(\Phi'\Phi)^{-1}\Phi'y. \tag{5.26}$$

When discrete Legendre polynomials are used, $\Phi'\Phi$ will be an identity matrix, and (5.26) simplifies. We establish the asymptotic properties of the 2SLS estimator in the following section. Further, we also show that the Sargan test, originally proposed as a test for overidentifying restrictions, can be used to examine whether or not y_t and x_t share a common trend.

We first give some assumptions before establishing the asymptotic results.

Assumption 5.3.1. Let $\{y_t, x_t\}_{t=1}^T$ be nonstationary trending time series. There exists a $(k+1)$-dimensional vector $\gamma_0 = (\alpha_0, \beta_0')'$ such that $e_t = y_t - \alpha_0 - x_t'\beta_0$ is strictly stationary with $\mathrm{E}[e_1] = 0$ and $\sigma_e^2 = \mathrm{E}[e_1^2] < \infty$.

Assumption 5.3.2. Let $\phi_j(\tau)$ be an orthogonal series of $\tau \in [0,1]$, for $1 \le j \le (p+1)$, such that $\phi_0(\tau_t) \equiv 1$, for $1 \le t \le T$, and:

(i) $\int_0^1 \phi_i(\tau)\phi_j(\tau)' d\tau = \delta_{ij}$, where $\delta_{ij} = 1$ for $i = j$ and $\delta_{ij} = 0$ for $i \ne j$. In addition, $\int_0^1 \phi_i(\tau)^4 d\tau < \infty$, for $i = 0, 1, 2, \ldots, p$;
(ii) $E[U'\varepsilon] = 0$, where $U = W^{1/2}\Phi' X$ and $\varepsilon = W^{1/2}\Phi' e$, with W defined as $W = (\Phi'\Phi)^{-1}$;
(iii) there exists a $(k+1)$-dimensional positive definite deterministic matrix D_T, with $\|D_T\|_2^2 \to \infty$ as $T \to \infty$, such that $\left(D_T^{-1}U'UD_T^{-1}, D_T^{-1}U'\varepsilon\right) \to_{\mathcal{D}} (\xi, \eta)$, where ξ is a matrix of positive random variables and η is a $(k+1)$-dimensional mixture normal random variable $\mathcal{N}(0, \xi\, \lambda_e^2)$ with $E\,||\xi||_2 < \infty$, in which $\lambda_e^2 = \mathrm{E}[e_1^2] + 2\sum_{j=1}^{\infty} E\left[e_1 e_{1+j}\right]$ is the long-run variance and $\|A\|_2 = \sqrt{\sum_{i=1}^d \sum_{j=1}^d a_{ij}^2}$ for matrix $A = \{a_{ij} : 1 \le i, j \le d\}$.

Remarks. (a) Assumption 5.3.1 establishes a co-trending relationship between the trending time series y_t and x_t and the coefficient vector β_0 in the linear combination $y_t - x_t'\beta_0$ that cancels the trends. (b) Assumption 5.3.2(i) defines a system of orthogonal basis functions. For example, it could be the first p *Legendre polynomials* normalized on $[0,1]$. (c) Assumption 5.3.2(ii) requires that the introduction of the Φ matrix can ensure that U and ε are uncorrelated. (d) Assumption 5.3.2(iii) imposes a high-level condition to ensure joint convergence. This can be verified under some low-level conditions imposed on (x_t, e_t) (such as Assumption 2.1 of Park and Phillips (2001) for the integrated time series case). Choi (2017) imposes a set of similar conditions for the pure unit-root setting. For more general trends, Lemmas B.1 and B.2 of Dong and Gao (2019) justify Assumption 5.3.2(iii). In Assumption 5.3.2(iii), one may choose D_T as a diagonal matrix to reflect the trending strength of x_{it}, for $i = 1, 2, \ldots, k$.

To illustrate that Assumption 5.3.2(iii) is a reasonable assumption, we consider a case where $k = 1$ and $W = I_p$. In this case, $U'U$ and $U'\varepsilon$ are given as follows:

$$U'U = \sum_{i=0}^{p} \begin{pmatrix} \left(\sum_{t=1}^T \phi_i(\tau_t)\right)^2 & \sum_{t=1}^T \phi_i(\tau_t) \sum_{t=1}^T \phi_i(\tau_t) x_t \\ \sum_{t=1}^T \phi_i(\tau_t) \sum_{t=1}^T \phi_i(\tau_t) x_t & \left(\sum_{t=1}^T \phi_i(\tau_t) x_t\right)^2 \end{pmatrix}$$

and

$$U'\varepsilon = \sum_{i=0}^{p} \begin{pmatrix} \sum_{t=1}^{T} \phi_i(\tau_t) \sum_{t=1}^{T} \phi_i(\tau_t) e_t \\ \sum_{t=1}^{T} \phi_i(\tau_t) x_t \sum_{t=1}^{T} \phi_i(\tau_t) e_t \end{pmatrix}.$$

In this situation, for each given i: (1) when $x_t = g(\tau_t) + v_t$, with $v_t \overset{i.i.d.}{\sim} \mathcal{N}(0,1)$ and $g(\tau)$ being a vector of unknown integrable functions of $\tau \in [0,1]$, we have $T^{-1}\sum_{t=1}^{T} x_t \phi_i(\tau_t) \rightarrow_P \int_0^1 g(r)\phi_i(r)dr$; (2) when $x_t = x_{t-1} + v_t$, with $v_t \overset{i.i.d.}{\sim} \mathcal{N}(0,1)$, we have $T^{-1}\sum_{t=1}^{T} x_t \phi_i(\tau_t) \rightarrow_D \int_0^1 B(r)\phi_i(r)dr$, where $B(\cdot)$ is a standard Brownian motion; and (3) when $x_t = x_{t-1} + g(\tau_t) + v_t$, with $v_t \overset{i.i.d.}{\sim} \mathcal{N}(0,1)$, we have $T^{-1}\sum_{t=1}^{T} \frac{x_t}{T}\phi_i(\tau_t) \rightarrow_P \int_0^1 \int_0^v g(u)\phi_i(v)dudv$. This shows that Assumption 5.3.2(iii) is satisfied when the trend in variables is nonlinear deterministic, stochastic, or a combination of the two. We now state the main theoretical results of the proposed method.

Theorem 5.3.1. *Let Assumptions* 5.3.1 *and* 5.3.2 *hold. As* $T \rightarrow \infty$, *the two-stage least squares estimator satisfies*

$$(U'U)^{1/2} (\widehat{\gamma}_{2sls} - \gamma_0) \rightarrow_D \mathcal{N}(0, \lambda_e^2 I_{(k+1)}), \tag{5.27}$$

where $U = W^{1/2}\Phi' X$ *and* λ_e^2 *is the long-run variance of* e_t.

Proof. See Appendix.

In practice, the long-run variance λ_e^2 can be estimated using a consistent estimator, such as the one proposed by Newey and West (1987, 1994) and Andrews (1991):

$$\widehat{\lambda}_e^2 = \sum_{j=-q_T}^{q_T} k(j/p_T)\widehat{\gamma}(j), \tag{5.28}$$

where $\widehat{\gamma}(j) = T^{-1}\sum_{t=1+j}^{T} \widehat{e}_t\widehat{e}_{t-j}$, in which $\widehat{e}_t = y_t - \widehat{\alpha}_{2sls} - x_t'\widehat{\beta}_{2sls}$. The truncation parameter q_T satisfies $q_T \rightarrow \infty$ and $q_T/T \rightarrow 0$ as $T \rightarrow \infty$. The kernel function $k(u)$ is defined on $[-1,1]$, and the Bartlett kernel $k(u) = (1 - |u|)\mathbf{1}(|u| < 1)$, in which $\mathbf{1}(\cdot)$ is the indicator function, is commonly used.

Remark. We note that the rate of convergence of the estimator depends on matrix D_T, ensuring $D_T^{-1}U'UD_T^{-1} \rightarrow_D \xi$.

Once we have obtained an estimate for γ_0, we need to test whether it is a true co-trending parameter. In other words, based on the estimated value of γ_0, we are interested in testing the following hypotheses:

$$\mathbb{H}_0: \quad x_t \text{ and } y_t \text{ share a common trend;} \tag{5.29}$$

$$\mathbb{H}_1: \quad x_t \text{ and } y_t \text{ do not share a common trend.} \tag{5.30}$$

The hypotheses state that no trend shall be detected in e_t under $\mathbb{H}_0$. We borrow the idea of an overidentification test from the conventional IV/GMM estimation theory that examines the validity of instrumental variables. That is, the IVs should be uncorrelated with the regression errors. Here, we consider an auxiliary regression

$$\widehat{e}_t = \theta_0 + \theta_1\phi_1(\tau_t) + \theta_2\phi_2(\tau_t) + \cdots + \theta_p\phi_p(\tau_t) + w_t, \tag{5.31}$$

for $t = 1, 2, \ldots, T$, where w_t is the regression error. Hence, in equation (5.25), "*the IVs are uncorrelated with the errors*" is equivalent to "*there is no trend in the errors.*" In other words, under $\mathbb{H}_0$, none of $\theta_1, \theta_2, \ldots, \theta_p$ in equation (5.31) should be significant. Following Sargan (1958), the test statistic is defined as

$$J_n(\widehat{\gamma}_{2sls}) = \frac{\widehat{e}'\Phi(\Phi'\Phi)^{-1}\Phi'\widehat{e}}{\widehat{\lambda}_w^2}, \tag{5.32}$$

where $\widehat{\lambda}_w^2$, which can be defined in the same way as in equation (5.28), is a consistent estimator of the long-run variance of w_t, and it is also a consistent estimator of λ_e^2 under $\mathbb{H}_0$.

To establish Theorem 5.3.2, we need to strengthen Assumption 5.3.1 as follows.

Assumption 5.3.3. (i) Let Assumption 5.3.1 hold. (ii) Suppose that e_t is α-mixing with coefficient $\alpha(j)$, satisfying $\sum_{j=1}^{\infty} \alpha^{\frac{\delta}{2+\delta}}(j) < \infty$, for some $\delta > 0$ such that $E\left[|e_1|^{2+\delta}\right] < \infty$. (iii) Let Assumption 5.3.2(i) hold. (iv) There is a consistent estimator $\widehat{\lambda}_w^2$ of λ_e^2 such that $\widehat{\lambda}_w^2/\lambda_e^2 \to_P 1$ under $\mathbb{H}_0$.

Remark. Assumptions 5.3.3(i) and (ii) are quite standard. Assumption 5.3.3(iv) can be justified by using existing results, such as those in Andrews (1991).

Theorem 5.3.2. *(Common features test for common trend) Let Assumption* 5.3.3 *hold. Then, as* $T \to \infty$, *we have under* $\mathbb{H}_0$,

$$J_T(\widehat{\gamma}_{2sls}) \to_{\mathcal{D}} \chi^2_{p-k}. \tag{5.33}$$

Proof. See the technical notes.

5.4 Simulated Example

In this section, we examine the finite sample performance of our test for a common trend in Theorem 5.3.2. We consider three kinds of trends: deterministic, stochastic, and broken. In the first two cases, the data-generating process is formulated as

$$y_t = (1 - \lambda)g_{1t} + \lambda g_{2t} + e_{1t}, \tag{5.34}$$

$$x_t = g_{1t} + e_{2t}, \tag{5.35}$$

for $t = 1, 2, \ldots, T$, where g_{1t} and g_{2t} are trend components, e_{1t} and e_{2t} are stationary error terms, and λ determines the commonality of trends in x_t and y_t. As in Kim *et al.* (2020), the error terms are generated by

$$(e_{1t}, e_{2t})' = L(v_{1t}, v_{2t})' \quad \text{in which } LL' = \begin{pmatrix} 1 & \rho \\ \rho & 1 \end{pmatrix}, \tag{5.36}$$

the shocks $(v_{1t}, v_{2t})'$ follow a vector autoregressive process that given by $(v_{1t}, v_{2t})' = \alpha(v_{1t-1}, v_{2t-1})' + (\varepsilon_{1t}, \varepsilon_{2t})'$, where $(\varepsilon_{1t}, \varepsilon_{2t})' \overset{i.i.d.}{\sim} \mathcal{N}(0, (1 - \alpha^2)I_2)$. Note that ρ controls the levels of endogeneity and α determines the strength of autocorrelation in the error terms. In the simulation, we consider $(0, 0.3, 0.5)$ for the values of α, $(0, 0.5)$ for ρ, and $(0, 0.3, 0.5)$ for λ, and the sample sizes equal 200 and 400, respectively. The trend components are specified in three different ways, as follows:

Case 1 (Deterministic trends): We assume that both g_{1t} and g_{2t} are deterministic trends. Following the simulation examples in

Cai (2007), which consider trending time-varying coefficient time series models, we work with the deterministic trend given by $g_{1t} = 2\tau_t + exp(-16(\tau_t - 0.5)^2) - 1$, which incorporates both complexity and nonlinearity. We then introduce a simple trending component defined by $g_{2t} = 2\sin(1.5\pi\tau_t)$, which provides an extra term that generates the scenario of no common trend when $\lambda \neq 0$.

Case 2 (Stochastic trends): We generate both g_{1t} and g_{2t} from pure random walk processes so that $g_{it} = g_{it-1} + u_{it}$ with $g_{i0} = 0$ for $i = 1, 2$, where $u_{1t} \overset{i.i.d.}{\sim} \mathcal{N}(0, 0.25)$ and $u_{2t} \overset{i.i.d.}{\sim} \mathcal{N}(0, 1)$, for $t = 1, 2, \ldots, T$. This specification is very common in the unit-root literature.

Case 3 (Broken trends): We assume that both g_{1t} and g_{2t} are linear functions of time with a break. Following the data-generating process in Kim *et al.* (2020) as well as the bilinear temperature trend estimated in Gao and Hawthorne (2006), we assume that both x_t and y_t contain linear trends with a break at $0.3T$ and $0.3T + 0.5\lambda T$, respectively. Specifically, x_t and y_t are generated by

$$y_t = 1 - 0.01t + 0.03b_t(0.3T + 0.5\lambda T) + e_{1t}, \tag{5.37}$$

$$x_t = 1 - 0.01t + 0.03b_t(0.3T) + e_{2t}, \tag{5.38}$$

where $b_t(t_0) = (t - t_0)1(t > t_0)$ and t_0 is the break point. When $\lambda = 0$, the two time series share a common trend with the same breakpoint. When $\lambda \neq 0$, the breakpoints differ by more and more as λ increases. The error terms $(e_{1t}, e_{2t})'$ are generated as in the previous examples.

For different scenarios in each of the cases, we run 10,000 independent simulations. We report the probabilities of rejecting the null hypothesis of a common trend using a 5% significance level in Table 5.1 with respect to different choices of $(\lambda, \alpha, \rho, T)$.

Table 5.1 reports the size and power of the proposed common trend test. When x_t and y_t share a common trend (i.e., $\lambda = 0$), the size of the test is near the predetermined significance level for the three cases. The size inflation is mild when the autoregressive coefficient α for the error term $(e_{1t}, e_{2t})'$ increases to 0.3 or 0.5. The power of the test grows with the level of difference between the trends in x_t and y_t controlled by λ. With a higher value of λ, the test presents a stronger power to signal the existence of trend features in $y_t - x_t'\widehat{\beta}$, indicating that the trends in x_t and y_t are not common. We also

Table 5.1. Probabilities of rejecting the null hypothesis of a common trend under 5% significance level.

			Size of the test		Power of the test			
			$\lambda = 0$		$\lambda = 0.3$		$\lambda = 0.6$	
	ρ	α	$T = 200$	$T = 400$	$T = 200$	$T = 400$	$T = 200$	$T = 400$
Case 1		0	0.061	0.055	0.907	0.996	1.000	1.000
	0	0.3	0.065	0.058	0.679	0.925	0.999	1.000
		0.5	0.066	0.060	0.502	0.770	0.987	1.000
		0	0.060	0.050	0.980	1.000	1.000	1.000
	0.5	0.3	0.061	0.055	0.872	0.993	0.998	1.000
		0.5	0.065	0.056	0.699	0.930	0.973	0.999
Case 2		0	0.054	0.053	0.919	0.969	0.980	0.994
	0	0.3	0.059	0.054	0.859	0.946	0.955	0.985
		0.5	0.061	0.056	0.790	0.908	0.927	0.967
		0	0.056	0.052	0.931	0.975	0.982	0.994
	0.5	0.3	0.060	0.054	0.884	0.951	0.957	0.982
		0.5	0.061	0.055	0.824	0.910	0.927	0.963
Case 3		0	0.059	0.054	0.805	1.000	0.998	1.000
	0	0.3	0.063	0.056	0.554	1.000	0.937	1.000
		0.5	0.069	0.057	0.500	1.000	0.789	1.000
		0	0.060	0.052	0.950	1.000	0.999	1.000
	0.5	0.3	0.061	0.056	0.795	1.000	0.963	1.000
		0.5	0.064	0.060	0.598	1.000	0.843	1.000

note that the performance of the test is not affected by the level of endogeneity, as the rejection rates for $\rho = 0.5$ are very similar to those for $\rho = 0$.

5.5 Summary

In this chapter, we have proposed a common features approach to test for common trends and estimate the co-trending relationship between trending time series. This approach does not rely on pretests to determine the exact nature of the trend in time series, and it is equally applicable to cases where variables have stochastic or nonlinear deterministic trends.

It is important to explain the difference between what we do here and what is reported by Müller and Watson (2018), who examined long-run covariability using growth rates of economic variables. In this chapter, we consider trends in the levels of the nonstationary trending variables. They examine variability at lower than business cycle frequencies in the growth rates of variables.

Specifically, let x_t be a time series such that Δx_t has significant variation at frequencies lower than the business cycle frequency. Consider $y_t = \alpha + \beta t + \gamma x_t + \varepsilon_t$, where α, β and γ are fixed constants and ε is white noise. It is obvious that y_t and x_t do not have a common trend. The reason is that any linear combination of the two still contain a trend component. However, Δy_t and Δx_t have significant long-run covariability. It is possible to define variation at lower than business cycle frequencies as a feature and develop a common feature-based test for covariability; however, that is outside of the interest of this chapter.

5.6 Technical Notes

Proof of Theorem 5.3.1. Note that the two-stage least squares estimator is established as

$$\widehat{\gamma}_{2sls} = (X'\Phi(\Phi'\Phi)^{-1}\Phi'X)^{-1}X'\Phi(\Phi'\Phi)^{-1}\Phi'y.$$

Replacing y using $y = X\beta + e$, we have

$$\widehat{\gamma}_{2sls} - \gamma_0 = (X'\Phi(\Phi'\Phi)^{-1}\Phi'X)^{-1}X'\Phi(\Phi'\Phi)^{-1}\Phi'e, \tag{5.39}$$

which, along with Assumption 5.3.2(iii), implies

$$\begin{aligned} &(X'\Phi(\Phi'\Phi)^{-1}\Phi'X)^{1/2}\left(\widehat{\gamma}_{2sls} - \gamma_0\right) \\ &= (X'\Phi(\Phi'\Phi)^{-1}\Phi'X)^{-1/2}X'\Phi(\Phi'\Phi)^{-1}\Phi'e \\ &\to_{\mathcal{D}} \mathcal{N}(0, \lambda_e^2\, I_{(k+1)}), \end{aligned} \tag{5.40}$$

by using the continuous mapping theorem (see, for example, Billingsley, 1999). This completes the proof of Theorem 5.3.1.

Proof of Theorem 5.3.2. Note that the residuals are defined as $\widehat{e} = y - X\widehat{\gamma}_{2sls}$ and the two-stage least squares estimator $\widehat{\gamma}_{2sls} - \gamma_0 =$

$\left(X'\Phi(\Phi'\Phi)^{-1}\Phi'X\right)^{-1}X'\Phi(\Phi'\Phi)^{-1}\Phi'e$. We then have

$$\begin{aligned}
(\Phi'\Phi)^{-1/2}\Phi'\widehat{e} &= (\Phi'\Phi)^{-1/2}\Phi'(y - X\widehat{\gamma}_{2sls}) \\
&= (\Phi'\Phi)^{-1/2}\Phi'(y - X\gamma_0) - (\Phi'\Phi)^{-1/2}\Phi'X(\widehat{\gamma}_{2sls} - \gamma_0) \\
&= \left[I - (\Phi'\Phi)^{-1/2}\Phi'X\left(X'\Phi(\Phi'\Phi)^{-1}\Phi'X\right)^{-1}X'\Phi(\Phi'\Phi)^{-1/2}\right] \\
&\quad \times (\Phi'\Phi)^{-1/2}\Phi'e \\
&= \widehat{M_1}(T)\widehat{M_2}(T),
\end{aligned}$$

where $\widehat{M_1}(T)=I-(\Phi'\Phi)^{-1/2}\Phi'X\left(X'\Phi(\Phi'\Phi)^{-1}\Phi'X\right)^{-1}X'\Phi(\Phi'\Phi)^{-1/2}$ and $\widehat{M_2}(T) = (\Phi'\Phi)^{-1/2}\Phi'e$. Under Assumption 5.3.3, similar to the proof of Theorem 2.21(i) by Fan and Yao (2003), as $T \to \infty$, we have

$$\widehat{M_2}(T) \to_{\mathcal{D}} \mathcal{N}(0, \lambda_e^2\ I_{(p+1)}).$$

Meanwhile, $\widehat{M_1}(T)$ is an idempotent $(p+1) \times (p+1)$ symmetric matrix with rank $p - k$ since

$$\begin{aligned}
\mathrm{tr}(M_1(T)) &= \mathrm{tr}(I_{p+1}) - \mathrm{tr}((\Phi'\Phi)^{-1/2}\Phi'X\left(X'\Phi(\Phi'\Phi)^{-1}\Phi'X\right)^{-1} \\
&\quad \times X'\Phi(\Phi'\Phi)^{-1/2}) \\
&= \mathrm{tr}(I_{p+1}) - \mathrm{tr}(X'\Phi(\Phi'\Phi)^{-1/2}(\Phi'\Phi)^{-1/2} \\
&\quad \times \Phi'X\left(X'\Phi(\Phi'\Phi)^{-1}\Phi'X\right)^{-1}) \\
&= \mathrm{tr}(I_{p+1}) - \mathrm{tr}(I_{k+1}) = p - k.
\end{aligned}$$

Also, under Assumption 5.3.3, $\widehat{\lambda}_w^2$ is a consistent estimator for λ_e^2. Therefore, based on the definition of $\widehat{M_1}(T)$ and $\widehat{M_2}(T)$, we have under $\mathbb{H}_0$,

$$J_T(\widehat{\gamma}_{2sls}) = \frac{\widehat{e}'\Phi(\Phi'\Phi)^{-1}\Phi'\widehat{e}}{\widehat{\lambda}_w^2} = \frac{\widehat{M_2}(T)'\widehat{M_1}(T)\widehat{M_2}(T)}{\widehat{\lambda}_w^2} \to_{\mathcal{D}} \chi_{p-k}^2.$$

The above result is obtained as $\widehat{M_2}(T)$ converges to a $(p+1)$-dimensional normal distribution, while $\widehat{M_1}(T)$ converges in probability to a $(k+1) \times (k+1)$-dimensional idempotent and symmetric matrix with rank $(p-k)$, and $\widehat{\lambda}_w^2$ converges to λ_e^2 in probability under $\mathbb{H}_0$. ∎

Chapter 6

Applications in Climate Change

> Climate change has caused substantial damages, and increasingly irreversible losses, in terrestrial, freshwater and coastal and open ocean marine ecosystems.... Widespread deterioration of ecosystem structure and function, resilience, and natural adaptive capacity, as well as shifts in seasonal timing, have occurred due to climate change, with adverse socioeconomic consequences.
>
> — IPCC Sixth Assessment Report

In recent decades, global temperatures have exhibited a warming trend, a phenomenon predominantly attributed to excess anthropogenic greenhouse gas emissions, as underscored by the Intergovernmental Panel on Climate Change report (IPCC, 2014). Scientists and international associations have been appealing for the control of global warming by reducing the emissions of greenhouse gases, which has led to much research in the area of climate change. Understanding the scientific mechanisms and economic consequences of global warming, including its impacts on agriculture, public health, sea level rise, ocean acidification, hurricane intensification, and ecosystem losses, has become a focal point in economic research (Nordhaus, 2013).

Climate econometrics is a new research area that as highlighted by Hsiang (2016), "identifying the effect of climate on societies is central to understanding historical economic development, designing modern policies that react to climatic events, and managing future global climate change". Castle and Hendry (2022) stated that "shared features of economic and climate time series imply that tools for empirically modeling nonstationary economic outcomes are also appropriate for studying many aspects of observational climate-change data". Our interest lies in delving into the trending behaviors of global temperatures and atmospheric concentrations of greenhouse gases, as well as the long-run relationship between them. Specifically, we aim to ascertain whether these variables share a common trend and to provide an estimation of the long-term sensitivity of global temperatures to greenhouse gas concentrations.

In this chapter, we first investigate the properties of the time series data of global temperatures and greenhouse gases and then establish the co-trending relationship between them based on the common features approach proposed in the previous chapter. The estimated climate sensitivity based on our approach is then compared with those in the literature. Moreover, we find that the cyclical patterns in temperatures are significant and non-negligible when conducting future forecasts of global temperatures conditional on various representative pathways of greenhouse gas emissions.

6.1 Global Temperatures and Greenhouse Gases

The upper panel of Figure 6.1 plots the annual Global Temperature Anomalies (GTAs) spanning from 1850 to 2018. GTA serves as a widely adopted metric for gauging global temperature levels, playing a crucial role in assessing climate change. We acquired the annual GTA time series from https://crudata.uea.ac.uk/cru/data/temperature/, a source providing both global and hemispheric combined land and ocean temperatures expressed as deviations from the 1960–1990 average. As depicted in the graph, an obvious upward warming trend is observable in GTA, particularly post the 1970s, coinciding with the heightened awareness of global warming and climate change in the recent thirty years. Preceding the 1970s, notable warming intervals occurred from 1910 to 1940 and again from 1975 to 2018, interspersed with periods of slight cooling from 1880 to 1910 and from 1940 to 1975. These trending patterns reflect the dynamic nature of global temperatures over the past 170 years.

The lower panel of Figure 6.1 presents the radiative forcing (measured in W/m^2) corresponding to concentrations of greenhouse gases (GHGs) considered in the Kyoto Protocol spanning from 1850 to 2018. These concentrations represent CO_2 equivalent levels

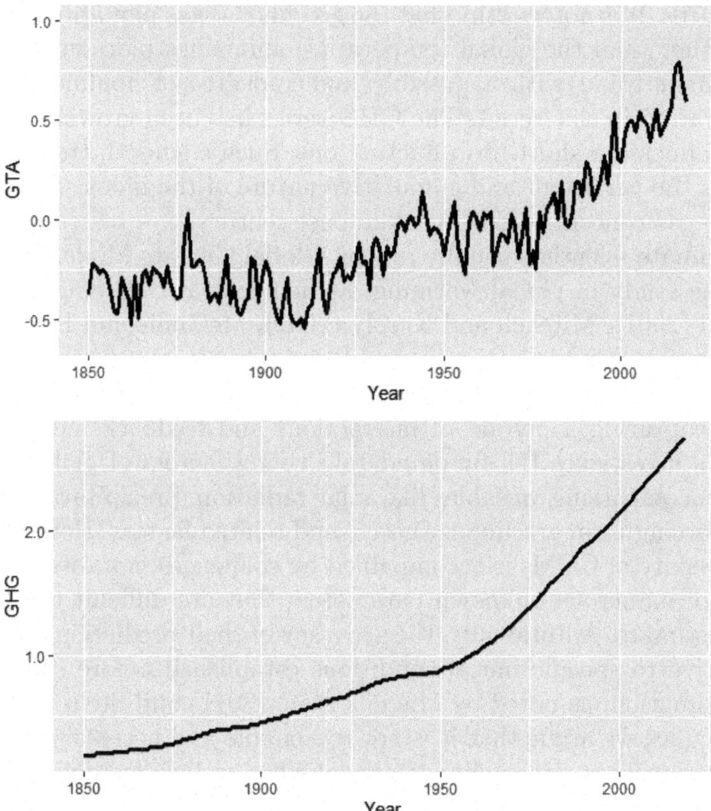

Figure 6.1. Global temperature anomalies and radiative forcing of greenhouse gases, 1850–2018.

of all anthropogenic forcings of greenhouse gases regulated under the Kyoto Protocol. The raw data, initially in parts per million (ppm), was obtained from https://iiasa.ac.at/models-tools-data/rcp. To convert this time series data of GHG concentrations into radiative forcings, we applied the formula $\text{GHG}_t = 5.35 \ln(C_t/C_0)$, where $C_0 = 278$ ppm serves as the reference concentration level and C_t denotes the concentration of greenhouse gases at time t.

Similar to the GTA, the GHG plot also reveals an upward trending trajectory. Notably, the GHG plot exhibits a steeper slope post-1950 compared to earlier periods. This pattern signifies a rapidly escalating concentration of greenhouse gases in the atmosphere, aligning with the rapid industrialization observed in numerous countries since the 1970s. It is noteworthy that the greenhouse gas plot appears much smoother than the global temperature anomalies plot, suggesting a dominant trend with a possibly higher degree of nonlinearity and integration in this series. The GHG series has been growing steadily with almost no short-term fluctuations. Such a smooth trend underscores the persistent and cumulative nature of the increase in greenhouse gas concentrations over the past century.

Climate scientists usually rely on Global Climate Models (GCMs) for the study of global warming, as demonstrated by Cubasch and Meehl (2001), Mitchell and Karoly (2001), McGuffie and Henderson-Sellers (2014), and Gettelman and Rood (2016), among others. These models are sophisticated representations of the intricate climate system, capturing a myriad of interactions and feedback mechanisms. GCMs investigate the fundamental driving factors of global warming, encompassing variables like solar radiation, precipitation, radiative forcing from greenhouse gases, and related factors. However, the complexity of GCMs is accompanied by challenges, notably the presence of numerous unknown parameters that are difficult to identify and estimate. Additionally, the accuracy of their predictions is highly sensitive to specific initial conditions established before calibration or estimation, as noted by Magnus *et al.* (2011) and Stein (2020). In particular, we argue that it is not reasonable and necessary to adopt an extremely complex climate model when we are merely interested in very few climate variables.

In the examination of the relationship between GTA and GHG, econometricians rely on time series models. By focusing on the key regression relationship of interest, the statistical models, in contrast to GCMs, are characterized by much fewer parameters, therefore resulting in a more concise framework. The methodology and implication of time series modeling depend on the form of trends in the variables. Therefore, econometricians have to make a foundational assumption about whether the trends in climate-related variables are deterministic or stochastic. Taking GTA as an example, this problem has sparked a protracted debate in the literature.

One group of researchers believes that the embedded trend in GTA is stochastic in that GTA is a unit-root process. As a consequence, shocks generate permanent effects on the temperature series. For example, Kaufmann and Stern (2002) and

Kaufmann *et al.* (2006) argued that GTA and radiative forcing of greenhouse gases contain stochastic trends (unit roots) that form a *cointegrating relationship.* Mills (2009) confirmed the robustness of such a relationship by examining the sensitivity of temperature to the radiative forcing of greenhouse gases with an updated dataset. Kaufmann *et al.* (2010) conducted an in-sample forecasting experiment to justify the rationality of viewing temperature series as an integrated process. Kaufmann *et al.* (2013) argued that physical mechanisms give rise to unit roots in radiative forcing of the atmospheric concentration of greenhouse gases and provide the source of unit roots in global temperatures. Recently, Chang *et al.* (2020) applied a new unit-root test which also supports the existence of a unit root in the temperature data.

When GTA and GHG are recognized as integrated (unit-root) time series that exhibit stochastic trends, a common practice is to use statistical tools to determine the orders of integration in both time series. In Table 6.1, unit root tests including the augmented Dickey–Fuller (ADF) test, the Dickey–Fuller GLS test by Elliott *et al.* (1996), and the KPSS test by Kwiatkowski *et al.* (1992) all indicate that the temperature series GTA is I(1), whereas the radiative forcing of the greenhouse gases series GHG is I(2) at the 5% significance level.

Specifically, when GTA is assumed to follow an I(1) process, we can obtain a fitted model using the data of GTA from 1850 to 2018 as

$$\underset{(s.e.)}{\Delta GTA_t} = \underset{(0.0089)}{0.0058} + \widehat{v}_t, \tag{6.1}$$

where $\widehat{v}_t$ is the estimated residuals of a weakly stationary process. The constant term is estimated as the sample mean of ΔGTA_t, with its standard error estimated by $sd(\Delta GTA_t)/\sqrt{T}$ in parentheses. The mean of ΔGTA_t measures the average warming speed of global temperature from 1850 to 2018, which, however, is not significant even at the 10% significance level. This implies that the trending characteristic in GTA is mostly dominated by a stochastic trend, and this embedded stochastic trend can be recovered by $GTA_t - 0.0058t - GTA_1$, for $t = 1, 2, \ldots, T$.

However, other econometricians believe that GTA is a trend-stationary time series that contains nonlinear deterministic trends. This means that temperature shocks only cause transitory impacts

Table 6.1. p-values of unit-root tests for GTA and GHG.

	ADF test	DF-GLS test	KPSS test
GTA	>0.10	>0.10	<0.01
ΔGTA	<0.01	<0.05	>0.10
GHG	>0.10	>0.10	<0.01
ΔGHG	>0.10	>0.10	<0.01
Δ^2GHG	<0.01	<0.01	>0.10

on the subsequent time series. For example, Gao and Hawthorne (2006) and Gay *et al.* (2009) argued that temperature series are better characterized by a trend-stationary process with a smooth nonlinear time trend, or a linear trend with breaks, rather than unit-root time series. Estrada *et al.* (2010) showed that econometric tests that allow for the possibility of any break in the deterministic trend in temperatures reject the hypothesis that temperature has a stochastic trend. As a result, the cointegration conclusion is not reliable because it is based on the assumption that historical climate data are integrated time series. Using data on hemispheric temperatures and a measure of aggregate radiative forcing with a shorter sample, Estrada *et al.* (2013) applied a nonlinear co-trending test proposed by Bierens (1997) to show that temperature and radiative forcing are stationary around piecewise linear trends, and they share a common nonlinear deterministic trend.

Therefore, we can also fit the GTA series by deterministic time trends. In Figure 6.2(a), we fit the GTA time series data by a linear time trend

$$\underset{(s.e.)}{\mathrm{GTA}_t} = -\underset{(0.0259)^{***}}{0.5264} + \underset{(0.0003)^{***}}{0.0051t} + \widehat{v}_{1t}, \tag{6.2}$$

which, however, is a spurious regression because the residuals $\widehat{v}_{1t}$ are found to contain a unit root.

In Figure 6.2(b), we adopt a piecewise linear trend that allows for only one trend break. The fitted model is

$$\underset{(s.e.)}{\mathrm{GTA}_t} = -\underset{(0.0213)^{***}}{0.4036} + \underset{(0.0003)^{***}}{0.0027t} + \underset{(0.0011)^{***}}{0.0136(t-t^*)1_{(t>t^*)}} + \widehat{v}_{2t}, \tag{6.3}$$

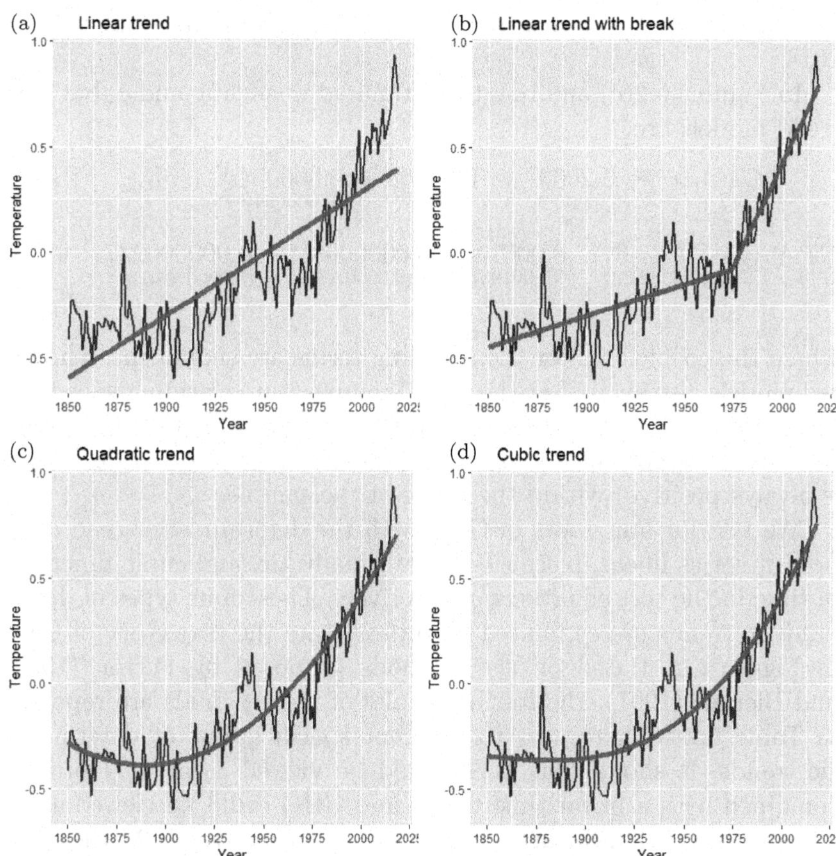

Figure 6.2. Global temperature anomalies fitted by linear, piecewise linear, quadratic, and cubic time trends.

where $1_{(\cdot)}$ is an indicator function, and the year of breakpoint is estimated as $t^* = 1974$ by a profile method. When examined by unit-root tests, the residuals $\widehat{v}_{2t}$ are stationary indicating that the piecewise linear time trend reasonably captures the trending pattern in GTA. The fitted model (6.3) shows that the average global temperature warming speed was 0.0027°C per year before 1974 and raised about six times to 0.0163°C per year after 1974. Hence, using a constant slope parameter (i.e., a linear trend) in Figure 6.2(a) is not appropriate because the warming speed is unlikely to be at a spurious invariant level of 0.0051°C per year over the 169 years from 1850 to 2018.

In Figures 6.2(c) and (d), the estimated quadratic and cubic time trend models are

$$\underset{(s.e.)}{GTA_t} = -\underset{(0.0287)^{***}}{0.2663} - \underset{(0.0008)^{***}}{0.0040t} + \underset{(0.000004)^{***}}{0.000054t^2} + \widehat{v}_{3t}, \tag{6.4}$$

$$\underset{(s.e.)}{GTA_t} = -\underset{(0.0382)^{***}}{0.3256} - \underset{(0.00193)}{0.00013t} - \underset{(0.0000265)}{0.0000068t^2} + \underset{(0.000000102)^{(***)}}{0.000000237t^3} + \widehat{v}_{4t}, \tag{6.5}$$

where the coefficients for their leading terms are significant. The fitted models also show that the global temperature warming speed has been rising as the first-order derivative of the estimated time trend function is monotonically increasing (i.e., the second-order derivative is always positive) within the range of the sample.

We test for unit roots in GTA with the deterministic trend component set as linear, piecewise linear, quadratic, and cubic functions of time in the test equations, respectively. These four types of deterministic trends are estimated to approximate the trajectory of GTA in Figure 6.2. Based on the methods proposed by Perron (1989) and Bierens (1997), the testing results of the residuals are reported in Table 6.2 with respect to different specifications of deterministic trends. It shows that GTA could be viewed as an I(1) process combined with a linear time trend, and GHG could be viewed as an I(2) process with a linear or even quadratic time trend. At the same time, however, when we make the deterministic trend component more nonlinear, the evidence of integration disappears for both time series. In other words, they could also be regarded as a nonlinear deterministic time trend with stationary errors.

Table 6.2. Specifications of deterministic trends and unit-root test results.

	Deterministic trend forms	Order of integration
GTA	$\alpha_0 + \alpha_1 t$	I(1)
	$\alpha_0 + \alpha_1 t + \alpha_2 t^2$	I(0)
	trend break at 1964	I(0)
GHG	$\alpha_0 + \alpha_1 t$	I(2)
	$\alpha_0 + \alpha_1 t + \alpha_2 t^2$	I(2)
	$\alpha_0 + \alpha_1 t + \alpha_2 t^2 + \alpha_3 t^3$	I(0)

The divergence in perspectives on the exact nature of trends in the GTA and GHG series can be attributed to the difference regarding the regression equations employed in unit-root tests. Specifically, in tests like the Augmented Dickey–Fuller (ADF) unit-root test, the deterministic trend is conventionally modeled as a linear function of time. While it is also reasonable to relax this assumption and consider nonlinear time trends, doing so introduces uncertainty regarding the existence of unit roots. In essence, it is plausible to consider both the GTA and GHG time series as stationary around a nonlinear deterministic trend. Our findings underscore the sensitivity of unit-root existence to the specification of the deterministic component in the alternative hypothesis. As established by Bierens (1997), allowing for nonlinear deterministic trends in the alternative hypothesis can render the determination of a unit root uncertain. This nuance in modeling choices causes difficulty in deciding the real nature of trends in the time series data like GTA and GHG.

From the view of data modeling, a unit-root structure or parametric functional forms of time, such as polynomials and those with trend breaks, may misspecify the complex trend component. To reveal the trend component in a much more flexible manner, the time trend can be assumed to be purely nonparametric:

$$\text{GTA}_t = f_t + v_{5t}, \tag{6.6}$$

where $f_t = f(\tau_t)$ is regulated as a smooth nonparametric function defined on $[0, 1]$. Since we do not impose any parametric forms on the time trend component, it is estimated by a purely data-driven approach, for example, the nonparametric local linear kernel estimation method. In particular, in the neighborhood of any local time point t', the trend component f_t can be approximated by a linear function of time $f_t \approx b_0 + b_1(t - t')$, where the unknown coefficient b_0 is $f_{t'}$ and b_1 is the first-order derivative $\partial f/\partial t|_{t=t'}$. At every local time t', the coefficient vectors $(b_0, b_1)'$ are estimated by minimizing the weighted sum of

$$(\widehat{b}_0, \widehat{b}_1)' = \arg\min_{b_0, b_1} \sum_{t=1}^{T} \left(\text{GTA}_t - b_0 - b_1(t - t')\right)^2 K_h((t - t')/T), \tag{6.7}$$

where $K_h(u) = K(u/h)/h$ in which $K(\cdot)$ is a kernel function and h is a bandwidth parameter that controls the smoothness of the estimated

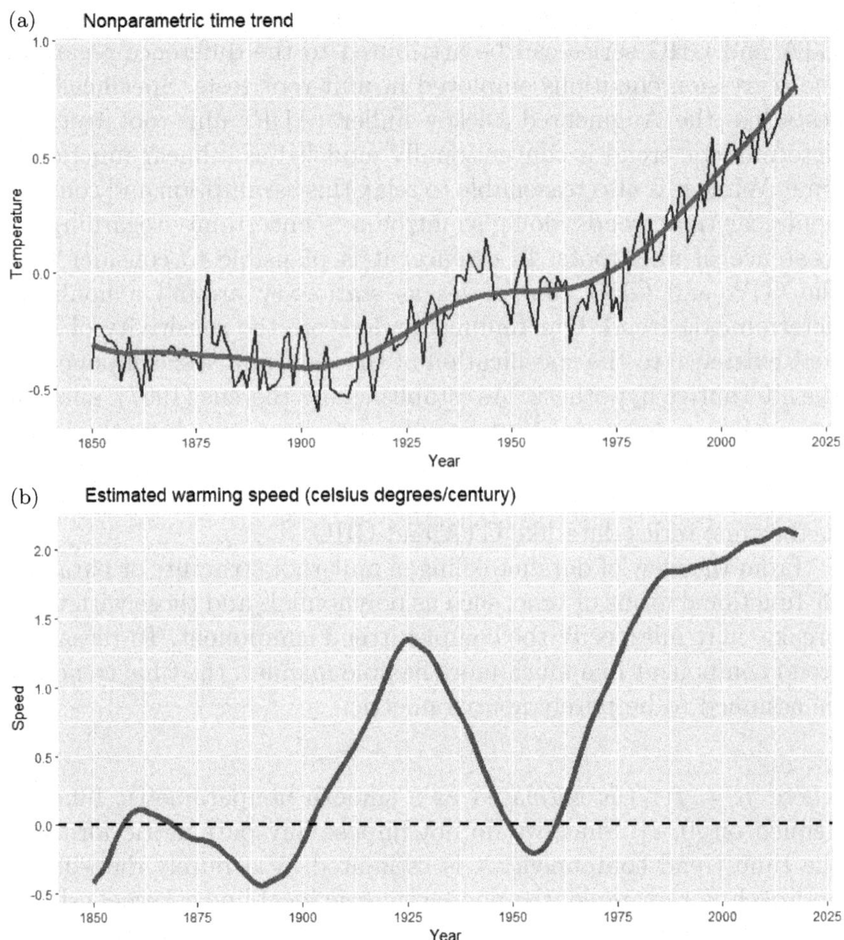

Figure 6.3. Nonparametric time trends in GTA.

time trend. Accordingly, the estimated time trend at t' is $\widehat{f}_{t'} = \widehat{b}_0$, while the estimated slope of time trend is $\partial \widehat{f}_{t'}/\partial t = \widehat{b}_1$.

Figure 6.3(a) presents the estimated nonparametric time trend using the Epanechnikov kernel $K(u) = 0.75(1 - u^2)1_{(|u|<1)}$ with bandwidth h set as 0.065. The estimated curve $\widehat{f}_t$ effectively captures the main evolving trajectory of GTA. Unit-root tests show that the residual sequence in equation (6.6) is stationary, suggesting that GTA_t can be reasonably modeled as a nonlinear trend-stationary process. In particular, the estimated deterministic trend is a flexible smooth function of time but not necessarily a polynomial of time. As a by-product of the local linear estimation method, Figure 6.3(b) plots the instantaneous speed of global warming. This is estimated as the first-order derivative of $\widehat{f}_t$ with respect to time, i.e., $\partial\widehat{f}_{t'}/\partial t = \widehat{b}_1$ at time t. Notably, Figure 6.3(b) is consistent with the scientific reality that the pace of global warming exhibits variations over time. Particularly noteworthy is the acceleration observed in recent decades, surpassing 2°C per 100 years after 2005. This acceleration poses a significant threat to the attainment of the objectives set in the *Paris Agreement*, whose central aim is "to strengthen the global response to the threat of climate change by keeping a global temperature rise this century well below 2°C above pre-industrial levels and to pursue efforts to limit the temperature increase even further to 1.5°C."

A preliminary finding from the empirical analysis indicates the formidable challenge of precisely characterizing the trends in the time series of GTA and GHG. These trends exhibit characteristics that can be attributed to stochastic trends stemming from integrated processes, yet they also display behaviors akin to deterministic trends represented by nonlinear functions of time. This complexity is similar to the longstanding debate on the nature of trends in macroeconomic variables, notably discussed in the work of Nelson and Plosser (1982). This problematic issue has motivated a large volume of literature in the context of macroeconomic data (see the works of Bierens, 1997; DeJong *et al.*, 1992; Kwiatkowski *et al.*, 1992; Nelson and Plosser, 1982, among others) and of global temperature data in the work of Breusch and Vahid (2011). Phillips (2005, 2010) referred to this difficult issue as *the mysteries of trend* that *"no one understands trend, but everyone sees them in the data"*, which is the major challenge of trending time series econometrics.

Trends in GTA and GHG cannot be correctly specified, as their exact nature is unknown and possibly unknowable (see the work of Phillips, 2010). These seemingly conflicting results reflect the problem caused by attributing nonstationarity to either deterministic or stochastic trends. Either unit root or nonlinear deterministic time trend may be a kind of approximation of the underlying trend. Given the theories established by Phillips (1998), they are equally valid as both deterministic and stochastic trends can be represented by orthogonal basis functions for trend, like two sides of the same coin.

Hence, one can use the approximation that is more convenient for the problem at hand. To this end, we exploit the fact that the nonlinear deterministic approximation to the underlying trend provides us with a convenient set of useful instruments to represent trends, test for common trends, and estimate the co-trending relationship.

6.2 The Co-trending Relationship

Given the fact that we are unable to decide the exact trending patterns in GTA and GHG, we apply the common features approach proposed in the previous chapter to test whether global temperature (GTA) and total radiative forcing of concentrations of greenhouse gases (GHG) share a common trend and, if so, estimate their co-trending relationship. We establish a coordinate system for trends using *Legendre polynomials* $\{\phi_j(\tau_t)\}_{j=0}^{p}$, with $\tau_t = t/T$ for $p \geq 2$ and $t = 1, 2, \ldots, T$, and take them as instruments to estimate the common trend coefficient in the regression model. We select $k = 3$ because Legendre polynomials $\phi_0(\tau_t)$ to $\phi_3(\tau_t)$ are sufficient to capture the trending patterns in the regressor GHG_t. The estimated co-trending relationship defined in equation (5.21) is presented in the following with estimated standard errors in parentheses:

$$\underset{(s.e.)}{\text{GTA}_t} = \underset{(0.035)^{***}}{-0.468} + \underset{(0.028)^{***}}{0.380}\ \text{GHG}_t + \widehat{e}_t. \tag{6.8}$$

The value of the test statistic in equation (5.32) for common trends (the Sargan test for overidentifying restrictions) is 2.195, which is less than 5.991, the critical value of χ^2_2 at the 5% significance level. This means that trends in GTA and GHG, whether stochastic or deterministic, are eliminated in the linear combination $\text{GTA}_t - 0.380\text{GHG}_t$. Therefore, they share a common trend and no trending patterns are detected in the residuals $\widehat{e}_t$. Furthermore, we also find that an AR(1) model is sufficient to characterize the dynamics in the residuals:

$$\widehat{e}_t = \underset{(0.063)^{(***)}}{0.577}\ \widehat{e}_{t-1} + \widehat{v}_t, \tag{6.9}$$

where no serial correlations remain in the sequence of $\widehat{v}_t$. Therefore, we conclude that GTA and GHG share a common trend. The estimation results for other choices of k are reported in Table 6.3.

Table 6.3. Estimation results of the co-trending relationship.

k	2	3	4	5
β_0	−0.4689	−0.4683	−0.4684	−0.4667
$(s.e.)$	(0.0346)	(0.0345)	(0.0345)	(0.0345)
β_1	0.3811	0.3804	0.3806	0.3788
$(s.e.)$	(0.0283)	(0.0281)	(0.0281)	(0.0281)
J-stat	2.139	2.195	3.753	6.489
$q_{0.95}(\chi^2_{k-1})$	3.841	5.991	7.815	9.488

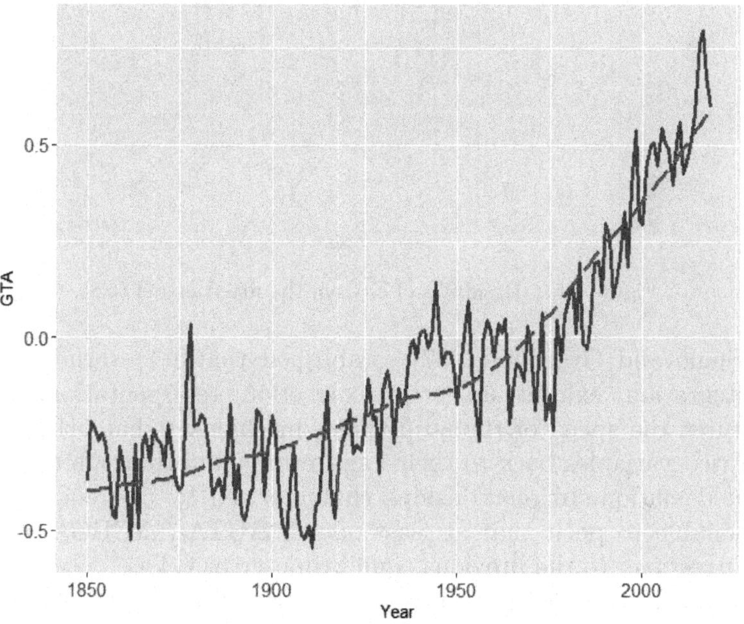

Figure 6.4. GTA and its fitted values by GHG.

The coefficient estimates are relatively stable regardless of the choice of k, and we always fail to reject the null hypothesis of common trend between GTA and GHG as the J-statistic is smaller than the corresponding critical value of the χ^2 distribution. The fitted values of GTA based on equation (6.8) and corresponding residuals are plotted in Figures 6.4 and 6.5.

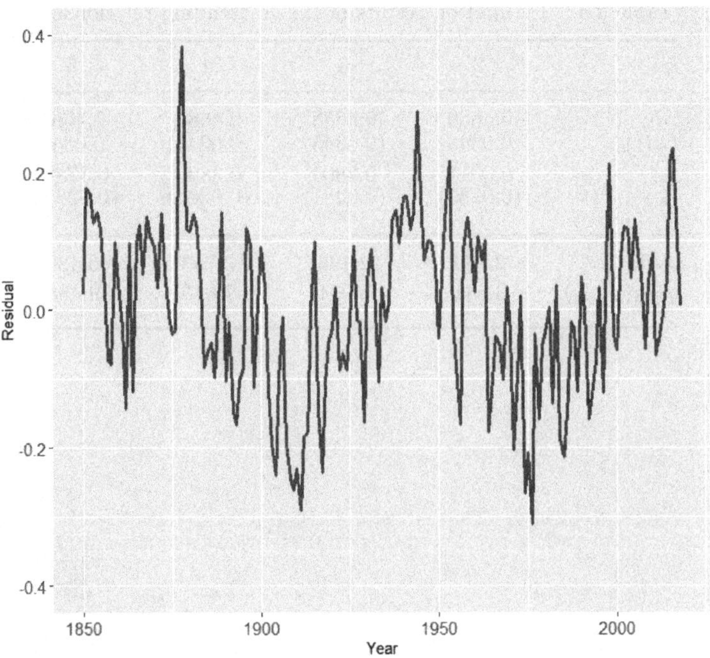

Figure 6.5. Residuals of $\widehat{e}_t$ from the fitted model (6.8).

Engle and Granger (1987) established that I(1) variables with cointegration exhibit an error correction representation, which captures the speed of the adjustment mechanism that brings cointegrated variables back to their long-term equilibrium after a short-term deviation. In relationships characterized by co-trending, it is reasonable to posit that at least one of ΔGTA_t, ΔGHG_t, or both must respond to the previous equilibrium error $\text{GTA}_{t-1} - \beta\text{GHG}_{t-1}$ to maintain their long-run relationship in the short run. To explore this error correction mechanism, we estimate the following two equations:

$$\widehat{\Delta\text{GTA}}_t = \text{const} - \underset{(0.072)^{***}}{0.427}\ \widehat{e}_{t-1} + \text{controls}, \tag{6.10}$$

$$\widehat{\Delta\text{GHG}}_t = \text{const} - \underset{(0.006)}{0.008}\ \widehat{e}_{t-1} + \text{controls}, \tag{6.11}$$

where $\widehat{e}_{t-1} = \text{GTA}_{t-1} + 0.468 - 0.380\text{GHG}_{t-1}$, standard errors are reported in parentheses, and *controls* represent the lagged values of

ΔGTA_t and ΔGHG_t. The nonresponse of GHG to the equilibrium error serves as a necessary condition for the exogeneity of GHG. This implies that, based on the history of these aggregate time series, no statistically significant feedback from the concentration of greenhouse gases to temperature can be detected.

6.3 Climate Sensitivity and Climate Cycles

The estimated response of temperature to radiative forcing of greenhouse gases is $0.38K/(W/m^2)$, and this indicates an approximate 1.41°C increase[1] in temperature if the concentration of greenhouse gases were doubled in the atmosphere. Our estimate of the sensitivity of temperature to radiative forcing of greenhouse gases is similar to previous estimates of this parameter from observational data. For example, Kaufmann *et al.* (2006) reported an estimate of sensitivity to an aggregate radiative forcing that includes forcing from solar irradiance and sulphur emissions in addition to radiative forcing of greenhouse gases using data from 1860 to 1994. If we consider only the radiative forcing of greenhouse gases, we obtain $0.41K/(W/m^2)$. Mills (2009), following the work of Kaufmann *et al.* (2006) estimation strategy and including only the radiative forcing of greenhouse gases, estimates the climate sensitivity to greenhouse gas concentration to be 0.39 with an extended data set that covers 1850–2000 (column 3 of Table 3 in the work of Mills, 2009). Using data from 1850–2010 and a completely different view about the nature of trend in temperature and greenhouse gas concentration, Estrada *et al.* (2013) used several temperature series with raw and filtered radiative forcing of greenhouse gases and reported a range of estimates from 0.35 to 0.43 (reported in Table 13 in the online supplement to the work of Estrada *et al.*, 2013). Here, using data from 1850 to 2018, we obtain a similar estimate with an agnostic view about the nature of the trend in temperature and greenhouse gas concentrations. In our analysis, however, the nature of the trend is unknowable from a finite sample of observations. The advantage of our method is that the knowledge of the exact nature of the trend is not needed for establishing whether

[1]It is calculated by $0.38 \times 5.35 \times \ln(2) \approx 1.41$.

two variables have a common trend and, if so, then estimating their co-trending relationship.

Climate science has two measures of climate sensitivity: transient climate response (TCR) and equilibrium climate sensitivity (ECS). TCR is the response of temperature to a doubling of the concentration of CO_2 in the atmosphere at the time of doubling, and ECS is the change in temperature after sufficient time is allowed for the system to get to equilibrium following a doubling of the CO_2 concentration. It is debatable whether time series models estimated using 160 years of observed data on two or three variables deliver an estimate of TCR or ECS (or perhaps something in between). Our estimate, 1.41°C increase in temperature at the time of doubling of CO_2 concentration, is in the range of TCRs of climate models reported in the fifth assessment report of the IPCC (1.1–2.6 with a median of 1.8, see Table 9.5 in https://www.ipcc.ch/site/assets/uploads/2018/02/WG1AR5_Chapter09_FINAL.pdf).

In this section, we focus on only two variables (temperature and radiative forcing of greenhouse gas concentrations) from the entire climate system. That is, we integrate all other climate variables. Previously, it was thought that concentrating on these two variables was too much reduction to be useful because unit-root tests suggested that global temperatures were I(1), while radiative forcing of greenhouse gases was I(2), so they could not have a long-run relationship. We show that if we allow for a more general trend, a common feature approach indicates that global temperatures and greenhouse gases have a common trend. This approach also allows us to estimate the co-trending relationship without an assumption about the nature of the trend. The advantage of this is that we are able to compute conditional forecasts given in Figure 6.10.

There are several possible reasons why estimates of the sensitivity of aggregate temperature to radiative forcing of greenhouse gases from time series models are smaller than the ECS in the climate models. For example, as Magnus *et al.* (2011) and Storelvmo *et al.* (2016) had documented, these aggregate models do not appropriately control for confounding factors such as aerosols that have also increased during this period but have a dimming effect. Also, for the high-emissions like RCP 8.5, the estimates of sensitivity from historical data do not include nonlinear feedback mechanisms that may be triggered when the concentration of the greenhouse gases or aggregate temperatures exceed levels that have not been experienced during the 1850–2018 period, whereas climate models do include such mechanisms.

In the papers by Pretis (2020) and Bruns *et al.* (2020), the authors argued that, under certain assumptions, a three-variable linear dynamic system, comprising global surface temperature anomalies (the same as GTA above), aggregate radiative forcing, and a proxy for the deep ocean temperature, is sufficiently close to a two-component energy balance model to provide estimates of ECS and TCR. This model implies a long-run relationship between GTA and aggregate radiative forcing. The aggregate forcing includes forcing of greenhouse gas concentrations, water vapor, ozone, black carbon, land use changes, snow albedo changes, and aerosols. Our common features approach can be used to estimate the long-run relationship between GTA and aggregate forcing directly without assuming the linearity of the dynamical system.

Using data from 1850 to 2018 and the same instruments as before, we obtain the estimated model

$$\text{GTA}_t = \underset{(0.025)^{***}}{-0.225} + \underset{(0.025)^{***}}{0.405}\ \text{ARF}_t + \widehat{e}_t. \tag{6.12}$$

where ARF is the aggregate radiative forcing for 1850–2018, which can be downloaded from Makiko Sato and James Hansen's website. It is labeled "Net forcing" in http://www.columbia.edu/~mhs119/Burden/Table.A1.ann.txt. The test for overidentifying restrictions is not rejected, confirming that GTA and ARF have a common trend. Assuming that all components of the aggregate increase proportionately, the estimated parameter $0.405K/(W/m^2)$ translates to an increase of 1.50°C in temperature if the concentrations of greenhouse gases were doubled in the atmosphere.

We have focused on the long-term trending components in the levels of GTA and GHG. The instrumental variables, consisting of orthogonal Legendre polynomials, capture the low-frequency signals (i.e., secular trends) in both series and reveal the co-trending regression relationship between them. Are the regression results affected by slightly higher signals than trends such as medium-term cycles? In particular, we find that the plot of the residuals of equation (6.8) in Figure 6.5 shows a regular cyclical pattern. Therefore, we extend equation (6.8) by adding a sine function to reveal the medium-term cycles in the GTA series. Using the nonlinear least squares estimation method, we have

$$\underset{(s.e.)}{\widehat{\text{GTA}}_t} = \underset{(0.035)^{***}}{-0.468} + \underset{(0.028)^{***}}{0.380}\ \text{GHG}_t + \underset{(0.011)^{***}}{0.106} \sin(\underset{(0.114)^{***}}{4.617}\pi\ \tau_t - \underset{(0.204)}{0.292}) + \widehat{\varepsilon}_t, \tag{6.13}$$

where the estimated cyclical sine function corresponds to a period of 73 years, with a recent local peak in 2016. The fitted sine function is plotted in Figure 6.6. After including the sine function component in the main regression equation, the new residuals $\widehat{\varepsilon}_t$ can be adequately fitted by an AR(1) model as well. In particular, we have

$$\widehat{\varepsilon}_t = \underset{(0.072)^{(***)}}{0.348}\,\widehat{\varepsilon}_{t-1} + \check{\varepsilon}_t, \tag{6.14}$$

where no serial correlation exists in $\check{\varepsilon}_t$. Note that the autoregressive coefficient estimate in equation (6.9) is larger than that in (6.14). A possible reason is that we may have mistakenly viewed the medium-term cycles as part of the serial correlations in the temperature series. It is worth noting that the amplitude of the estimated cyclical component is approximately 0.106°C, which is not only significant at the 1% level but also non-negligible compared to the variations in the GTA. In other words, medium-term cycles may also play an important role in the process of global warming we are experiencing today.

Figure 6.7 shows that if we include the medium-term cyclical component in the regression model, the newly fitted values of GTA fluctuate around those by only GHG. These cycles reinforce the warming effects caused by GHG when the sine function takes positive values, for example, during the periods from 1853 to 1889, from 1925 to 1962, and from 1998 until now. On the other hand, they offset the greenhouse gas effects when they turn to the negative side, during the periods from 1889 to 1925, and from 1962 to 1998. As a result, ignoring the medium-term cyclical component may misinterpret the trending behaviors in local periods. For example, cooling periods from 1875 to 1900 are fitted as mild warming periods by only GHG. We may also underestimate the warming speed from 1975 onwards if we neglect the 73-year medium-term cycle in GTA. Table 6.4 gives the estimated effects caused by GHG and cycle during each of the upward and downward phases of the sine function.

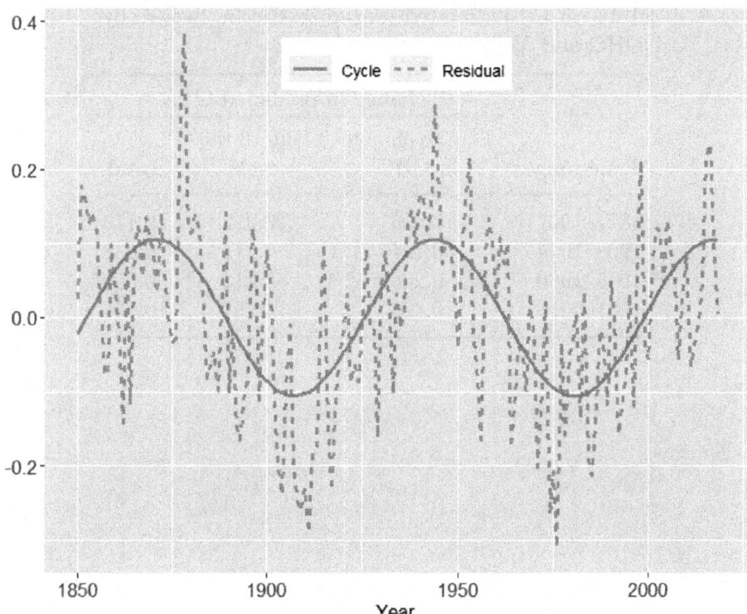

Figure 6.6. Cycles in the residuals.

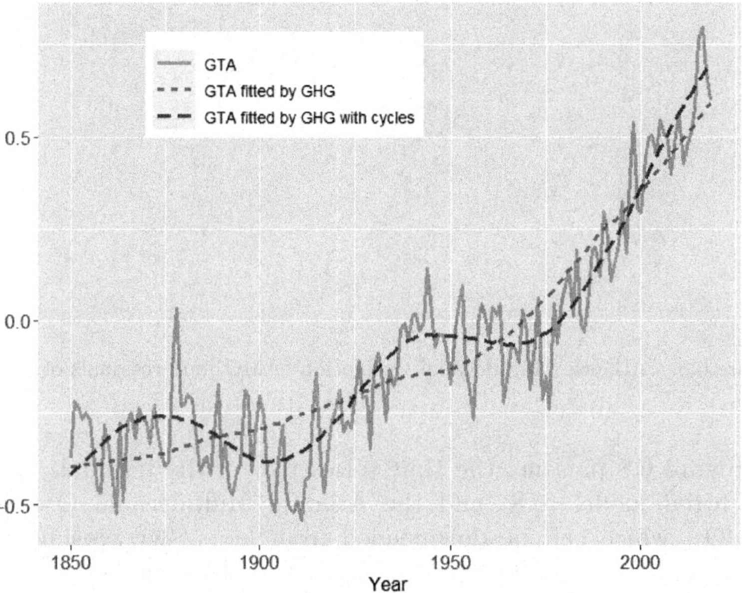

Figure 6.7. Fitted values of GTA with and without cycles.

Table 6.4. Estimated warming effects caused by GHG and medium-term cycles.

	Change in predicted GTA		
Time range	Due to change in GHG	Due to the cycle	Total
1871–1907	0.096	–0.212	–0.116
1908–1944	0.121	0.212	0.333
1945–1980	0.268	–0.212	0.056
1981–2016	0.426	0.212	0.638

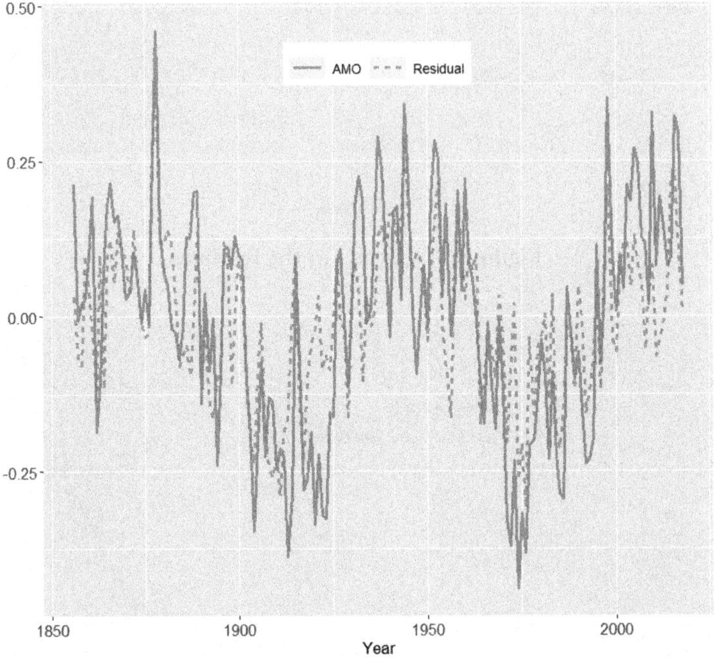

Figure 6.8. Atlantic Multidecadal Oscillation (AMO) and residuals of equation (6.8).

Figure 6.8 presents the time series plot of the residuals $\widehat{e}_t$ from the fitted model (6.8) and the Atlantic Multidecadal Oscillation (AMO), which can be downloaded from http://www.psl.noaa.gov /data/timeseries/AMO, and the scatter plot of the two series are given in Figure 6.9. Both figures show that the cyclical pattern in the residuals is in sync with the AMO. The AMO is a climate

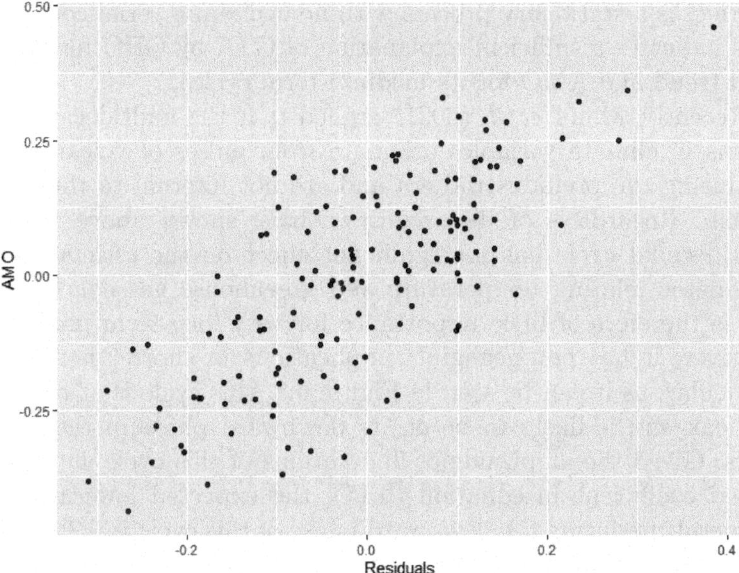

Figure 6.9. Scatter plot of the Atlantic Multidecadal Oscillation (AMO) and residuals of equation (6.8).

phenomenon characterized by long-term variations in sea surface temperatures (SSTs) in the North Atlantic Ocean. It is a natural climate cycle and operates independently of human-induced climate change. Several climate scientists have documented a cycle with a period of 60–80 years using 8000 years of climate proxy records (corals, tree rings, and ice cores) (see, for example, the work of Knudsen *et al.*, 2011). Estrada *et al.* (2013) discuss the possibility that the long cycle in AMO could distort the estimation of climate sensitivity parameters from time series data and they remove its influence from the temperature series at the outset. Adding AMO as an exogenous variable to equation (6.8), we obtain

$$\underset{(s.e.)}{\mathrm{GTA}_t} = -\underset{(0.011)^{***}}{0.469} + \underset{(0.009)^{***}}{0.375}\,\mathrm{GHG}_t + \underset{(0.036)^{***}}{0.515}\,\mathrm{AMO}_t + \widehat{u}_t, \quad (6.15)$$

which produces a very similar estimate of the sensitivity parameter as in equation (6.8), but it removes the cyclical pattern from the residuals. Specifically, the residuals in model (6.15) can be modeled by

$$\widehat{u}_t = \underset{(0.073)^{***}}{0.359}\,\widehat{u}_{t-1} + \zeta_t, \quad (6.16)$$

where ζ_t is a stationary process with no cycles nor serial correlation. This indicates a sufficient explanation of GTA by GHG for its long-term trend and AMO for its medium-term cycles.

Recently, Mann *et al.* (2021) argued that the multidecadal oscillations in climate variables originate from pulses of volcanic activity during the preindustrial era and are not internal to the climate system. Regardless of its origin, we have shown above that this multidecadal cycle has no significant effect on the estimate of the parameter relating temperature and greenhouse gas concentration and is therefore of little importance for very long-term projections. However, it has non-negligible implications in shorter horizons. In particular, as it can be seen in Figure 6.8, this cycle is at or around its peak and is likely to be on its downward phase in the next 30 years. Given the amplitude of fluctuations of this cycle and its estimated coefficient in equation (6.15), the expected lowering of the temperature during the downward phase of this cycle is 0.2°C, which is half as much as the expected increase in temperature during this period due to GHG concentration conditional on RCP 6.0. For the next few decades, this could make it difficult for the general public to appreciate the full effect of greenhouse gases on global temperatures.

6.4 Conditional Forecasts of Future Temperatures

The estimation and testing results of equation (6.8) suggest a stable long-run relationship between GTA and radiative forcing of greenhouse gases. The forecasts of future temperatures are based on our model conditional on different representative concentration pathways (RCPs) of greenhouse gases. The RCPs are scenarios used in climate modeling and research to explore different possible future trajectories of greenhouse gas concentrations in the Earth's atmosphere. These pathways were developed by the scientific community as part of the Coupled Model Intercomparison Project Phase 5 (CMIP5), which contributed to the Fifth Assessment Report of the Intergovernmental Panel on Climate Change (IPCC).

The conditional forecasts are plotted in Figure 6.10 when medium-term cyclical components are not incorporated in the co-trending regression (6.8). The figure plots the path of future GTA under three scenarios: the catastrophic path conditional on RCP 8.5 without significant mitigation efforts, the medium-level path conditional on RCP

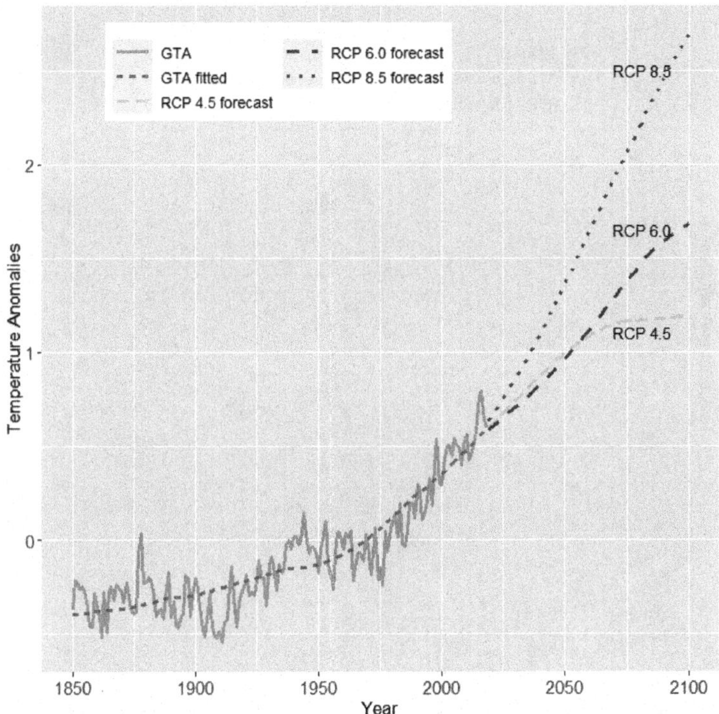

Figure 6.10. Conditional forecasts of future temperatures for different RCPs without cycles.

6.0 with intermediate levels of greenhouse gas emissions and mitigation efforts, and the low-level stabilization path conditional on RCP 4.5 with moderate mitigation efforts. Given that the co-trending relationship between GTA and GHG exists and remains unchanged in the future, the figure shows that under the worse scenario of RCP 8.5 of high greenhouse gas emissions, the global temperatures will rise continuously by about 2°C in 2100 compared with the level it is today.

When the 73-year cycle patterns are considered and incorporated as in model (6.13), future forecasts of GTA are plotted in Figure 6.11 for different RCPs. They are largely different from the conditional predictions in Figure 6.10 where cycles are not taken into account. As the cycle has entered a declining period since 2016, it offsets the greenhouse effect to some extent in the next 36 years. Therefore, the predicted paths of GTA under RCP 4.5 and RCP 6.0 are relatively flat until 2052. In other words, the global temperature is likely to

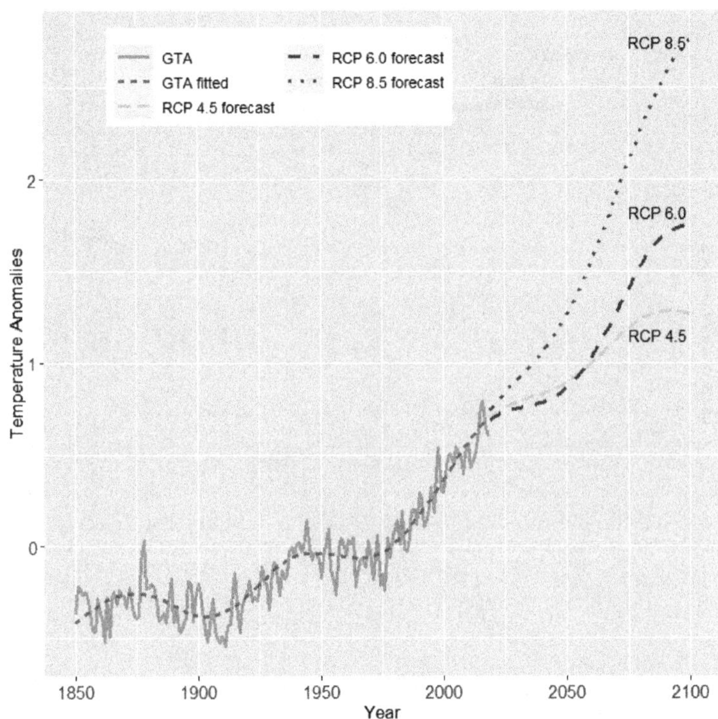

Figure 6.11. Conditional forecasts of future temperatures for different RCPs with cycles.

remain at a steady level if greenhouse gas emissions evolve at or below the medium-level path RCP 6.0. Therefore, we perceive the upcoming three decades as a critical window period for the implementation of climate change mitigation policies and the facilitation of international collaborations. Otherwise, the global temperatures will rise sharply under the scenario of RCP 8.5 and even RCP 6.0 after 2052.

Table 6.5 gives detailed statistics that measure predicted GTA changes due to GHG and medium-cycles under RCP 4.5, RCP 6.0, and RCP 8.5, respectively. The key message is that we may overestimate (underestimate) temperature raises during the downward (upward) phase of the medium-term cycle. For example, under the RCP 6.0 scenario, the downward cycle eliminates more than half of the GHG warming effect, causing only a 0.195°C increase over 33 years. When the two effects are combined in the rising cycle, the total increment is 0.797°C from 2053 to 2089, so temperatures will rise four times as much as the declining cycle.

Table 6.5. Changes in temperature forecasts, 2019–2089.

	ΔGTA conditional on RCP 4.5		
Time range	Due to ΔGHG	Due to the cycle	Total
2019–2052	0.418	–0.212	0.206
2053–2089	0.159	0.212	0.371
	ΔGTA conditional on RCP 6.0		
Time range	Due to ΔGHG	Due to the cycle	Total
2019–2052	0.407	–0.212	0.195
2053–2089	0.585	0.212	0.797
	ΔGTA conditional on RCP 8.5		
Time range	Due to ΔGHG	Due to the cycle	Total
2019–2052	0.781	–0.212	0.569
2053–2089	1.003	0.212	1.215

It is very important to note that our forecasts are conditional on RCPs and also conditional on all else remaining constant (i.e., remaining as they were during the estimation sample). This means that they are conditional on the feedback from temperature to greenhouse gas concentration remaining insignificant, as it has been during the estimation period. They are also conditional on no additional mechanisms starting to operate as temperatures rise. This *ceteris paribus* assumption may not be a good approximation if there are climate tipping points or new mechanisms that amplify the temperature increases. Hence, although prediction intervals are correct based on the "all else constant" assumption, they only reflect our confidence in these forecasts for the first few decades beyond the estimation sample.

6.5 Summary

In this chapter, we find that global temperatures and atmospheric concentrations of greenhouse gases share a common trend. Our estimate of the parameter linking the radiative forcing of greenhouse gases to global temperature is similar to the estimates published in previous studies that use observational data and within the range of transient climate response parameters of climate models used by the IPCC.

We document that the deviation of the global temperature series from its long-run equilibrium has a very long cycle with a nontrivial amplitude and is likely to have gone through its peak just recently. The downward phase of this cycle can potentially confuse the inferences made by casual observers in the near future. Countries with some emission control policies may infer overly optimistic conclusions about the success of their policies in reducing the warming trend in temperature. At the same time, countries with few or no emission control policies may conclude that their lack of action was warranted.

6.6 Bibliographical Notes

Climate change econometrics is a new research area. The work of Stern and Kaufmann (1999) was among the first group of papers studying climate time series data using econometric models. Pretis and Hendry (2013), Hsiang *et al.* (2013), Carleton and Hsiang (2016), Hsiang (2016), Stern (2016), and Castle and Hendry (2022) provided general introductions on the impacts of climate change on human society and on how econometric methods can be applied to analyze climate change problems. Econometricians have since investigated fundamental features of climate change econometrics. For example, Burke *et al.* (2015) found that the relationship between fundamental productive elements of modern economies and local temperature may be highly nonlinear. Pretis (2021) considered a full empirical climate-economic system and discussed the exogeneity conditions in climate econometrics.

In particular, trend analysis methods for climate time series was reviewed by Mudelsee (2019). Some important findings on climate trends include Lanzante (1996) for historical radiosonde station data, Easterling *et al.* (2000) regarding extreme climate events, Bengtsson *et al.* (2004) for reanalysis data, Hannachi (2007) for gridded climate data, Gil-Alana (2008) based on segmented trends and fractional integration methods, McKitrick *et al.* (2010) using panel and multivariate methods, Wang *et al.* (2012) using high-resolution spatial climate data, and Carson *et al.* (2019) for a review on climate model uncertainty and trend detection.

Appendix A

Proofs of the Lemmas

A.1 Proofs of the Lemmas in Chapter 3

Proof of Lemma 3.8.1.
Equation (3.52) in Lemma 3.8.1. Recall that $\tau_t = t/T$ and

$$M_1(i,j) = \frac{1}{T}\sum_{t=1}^{T}\left(g_i(\tau_t) - \bar{g}_{i,T}\right)\left(g_j(\tau_t) - \bar{g}_{j,T}\right), \tag{A.1}$$

where $\bar{g}_{i,n} = T^{-1}\sum_{t=1}^{T} g_i(\tau_t)$. Since $g(\cdot)$ is a continuous differentiable function of $\tau \in [0,1]$, under Assumption 3.3.1, we have

$$\bar{g}_{i,n} = \frac{1}{T}\sum_{t=1}^{T} g_i\left(\frac{t}{T}\right) \longrightarrow \int_0^1 g_i(\tau)d\tau = \bar{g}_i, \tag{A.2}$$

as the Riemann sum converges to its integral limit. The same argument applies to $M_1(i,j)$:

$$\begin{aligned}\frac{1}{T}\sum_{t=1}^{T}\left(g_i(\tau_t) - \bar{g}_{i,T}\right)\left(g_j(\tau_t) - \bar{g}_{j,T}\right) \longrightarrow \int_0^1 (g_i(\tau) - \bar{g}_i)(g_j(\tau) - \bar{g}_j)d\tau \\ = \Sigma_g(i,j).\end{aligned} \tag{A.3}$$

Therefore, as $T \to \infty$, $M_1(i,j) \longrightarrow \Sigma_g(i,j)$, for $i,j = 1,\ldots,k$.

Equation (3.53) in Lemma 3.8.1. Applying the Cauchy–Schwarz inequality, we have

$$
\begin{aligned}
|M_2(i,j)| &= \left| \frac{1}{T}\sum_{t=1}^{T}\left(\bar{g}_{i,T} - \sum_{s=1}^{T} w_{Ts}(t) g_i(\tau_s)\right)\left(\bar{g}_{j,T} - \sum_{s=1}^{T} w_{Ts}(t) g_j(\tau_s)\right)\right| \\
&\le \sqrt{\left|\frac{1}{T}\sum_{t=1}^{T}\left(\bar{g}_{i,T} - \sum_{s=1}^{T} w_{Ts}(t) g_i(\tau_s)\right)^2\right| \left|\frac{1}{T}\sum_{t=1}^{T}\left(\bar{g}_{j,T} - \sum_{s=1}^{T} w_{Ts}(t) g_j(\tau_s)\right)^2\right|}.
\end{aligned} \tag{A.4}
$$

Hence, we are able to show that $M_2(i,j) = o_p(1)$ if as $T \to \infty$,

$$
\frac{1}{T}\sum_{t=1}^{T}\left(\bar{g}_{i,T} - \sum_{s=1}^{T} w_{Ts}(t) g_i(\tau_s)\right)^2 = o_p(1), \tag{A.5}
$$

for any $i = 1, 2, \ldots, k$. Since (A.5) is always positive, we only need to show that

$$
\mathrm{E}\left[\frac{1}{T}\sum_{t=1}^{T}\left(\bar{g}_{i,T} - \sum_{s=1}^{T} w_{Ts}(t) g_i(\tau_s)\right)^2\right] = o(1). \tag{A.6}
$$

To prove the above result, we first write the equation as

$$
\begin{aligned}
&\frac{1}{T}\sum_{t=1}^{T}\left(\bar{g}_{i,T} - \sum_{s=1}^{T} w_{Ts}(t) g_i(\tau_s)\right)^2 \\
&\quad = \frac{1}{T}\sum_{t=1}^{T}\left(\sum_{s=1}^{T} w_{Ts}(t)\left(g_i(\tau_s) - \bar{g}_{i,T}\right)\right)^2 \\
&\quad = \frac{1}{T}\sum_{t=1}^{T}\sum_{s=1}^{T} w_{Ts}(t)^2 \left(g_i(\tau_s) - \bar{g}_{i,T}\right)^2 \\
&\qquad + \frac{1}{T}\sum_{t=1}^{T}\sum_{s_1=1}^{T}\sum_{\substack{s_2=1,\\ s_2\neq s_1}}^{T} w_{Ts_1}(t) w_{Ts_2}(t)\left(g_i(\tau_{s_1}) - \bar{g}_{i,T}\right)\left(g_i(\tau_{s_2}) - \bar{g}_{i,T}\right).
\end{aligned} \tag{A.7}
$$

Therefore, let $\xi_{s,t} = \frac{K\left(\frac{v_s - v_t}{h}\right)}{hf(v_t)}$, and it suffices to show that

$$\mathrm{E}[M_{2,i}(T)] = o(1), \tag{A.8}$$

for $i = 1, 2$, where

$$M_{2,1}(T) = \frac{1}{T^3} \sum_{t=1}^{T} \sum_{s=1}^{T} \xi_{s,t}^2 \left(g(\tau_s) - \bar{g}_T\right)^2 \tag{A.9}$$

and

$$M_{2,2}(T) = \frac{1}{T^3} \sum_{t=1}^{T} \sum_{s_1=1}^{T} \sum_{\substack{s_2=1,\\ s_2 \neq s_1}}^{T} \xi_{s_1,t} \xi_{s_2,t} \left(g(\tau_{s_1}) - \bar{g}_T\right) \left(g(\tau_{s_2}) - \bar{g}_T\right). \tag{A.10}$$

They directly imply that $M_2(i,j) = o_p(1)$.

Equations (3.54) and (3.55) in Lemma 3.8.1. Using the Cauchy–Schwarz inequality,

$$|M_{12}(i,j)| \leq |M_1(i,i)|^{1/2} |M_2(j,j)|^{1/2} = O_p(1) o_p(1) = o_p(1). \tag{A.11}$$

The same result holds for $M_{21}(i,j)$ for $i, j = 1, \ldots, k$.

Equation (3.56) in Lemma 3.8.1. For $S_2(i,j)$, using the Cauchy–Schwarz inequality, we have

$$\begin{aligned} |S_2(i,j)| &= \left| \frac{1}{T} \sum_{t=1}^{T} \left(v_{it} - \sum_{s=1}^{T} w_{Ts}(t) v_{is} \right) \left(v_{jt} - \sum_{s=1}^{T} w_{Ts}(t) v_{js} \right) \right| \\ &\leq \left| \frac{1}{T} \sum_{t=1}^{T} \left(v_{it} - \sum_{s=1}^{T} w_{Ts}(t) v_{is} \right)^2 \right|^{1/2} \left| \frac{1}{T} \sum_{t=1}^{T} \left(v_{jt} - \sum_{s=1}^{T} w_{Ts}(t) v_{js} \right)^2 \right|^{1/2}, \end{aligned} \tag{A.12}$$

for $i, j = 1, \ldots, k$, where

$$\begin{aligned} \frac{1}{T} \sum_{t=1}^{T} \left(v_{it} - \sum_{s=1}^{T} w_{Ts}(t) v_{is} \right)^2 &= \frac{1}{T} \sum_{t=1}^{T} \left(\sum_{s=1}^{T} w_{Ts}(t)(v_{is} - v_{it}) \right)^2 \\ &= \frac{1}{T} \sum_{t=1}^{T} \sum_{s=1}^{T} w_{Ts}(t)^2 (v_{is} - v_{it})^2 \\ &\quad + \frac{1}{T} \sum_{t=1}^{T} \sum_{s_1=1}^{T} \sum_{\substack{s_2=1,\\ s_2 \neq s_1}}^{T} w_{Ts_1}(t) w_{Ts_2}(t) (v_{is_1} - v_{it})(v_{is_2} - v_{it}). \end{aligned} \tag{A.13}$$

As in the previous proofs, we assume $k = 1$. As $\widehat{f}(v) = f(v) + o_p(1)$, it suffices to prove that (A.13) is $o_p(1)$. Since (A.13) is always positive, we only need to show that the expectations of the following two expressions are $o(1)$:

$$S_{2,1} = \frac{1}{T^3} \sum_{t=1}^{T} \sum_{\substack{s=1 \\ s \neq t}}^{T} \xi_{s,t}^2 (v_s - v_t)^2, \tag{A.14}$$

$$S_{2,2} = \frac{1}{T^3} \sum_{t=1}^{T} \sum_{\substack{s_1=1 \\ s_1 \neq t}}^{T} \sum_{\substack{s_2=1, \\ s_2 \neq t, s_2 \neq s_1}}^{T} \xi_{s_1,t} \xi_{s_2,t} (v_{s_1} - v_t)(v_{s_2} - v_t), \tag{A.15}$$

where $\xi_{s,t} = K\left(\frac{v_s - v_t}{h}\right) / h f(v_t)$. Thus, $S_{2,1} = o_p(1)$ and $S_{2,2} = o_p(1)$ imply $S_2(i,j) = o_p(1)$ for any $i, j = 1, 2, \ldots, k$.

Proof of Lemma 3.8.2. Without loss of generality, we let $k = 1$ so that

$$\begin{aligned} I_{1T} &= \frac{1}{\sqrt{T}} \sum_{t=1}^{T} \zeta(v_t) \widetilde{v}_t^2 = \frac{1}{\sqrt{T}} \sum_{t=1}^{T} \zeta(v_t) \left(\frac{\frac{1}{Th} \sum_{s=1}^{T} K\left(\frac{v_s - v_t}{h}\right)(v_s - v_t)}{\widehat{f}(v_t)} \right)^2 \\ &= \frac{(1 + o_p(1))}{\sqrt{T}} \sum_{t=1}^{T} \zeta(v_t) \left(\frac{\frac{1}{Th} \sum_{s=1}^{T} K\left(\frac{v_s - v_t}{h}\right)(v_s - v_t)}{f(v_t)} \right)^2 \\ &\triangleq (1 + o_p(1)) \widetilde{I}_{1T}. \end{aligned} \tag{A.16}$$

Hence, it is equivalent to show $\widetilde{I}_{1T} = o_p(1)$. We consider the square of $\widetilde{I}_{1T}$ as follows:

$$\begin{aligned} \widetilde{I}_{1T}^2 &= \frac{1}{T^5 h^4} \sum_{t_1=1}^{T} \sum_{t_2=1}^{T} \zeta(v_{t_1}) \zeta(v_{t_2}) \left(\frac{\sum_{s=1}^{T} K\left(\frac{v_s - v_{t_1}}{h}\right)(v_s - v_{t_1})}{f(v_{t_1})} \right)^2 \\ &\quad \times \left(\frac{\sum_{l=1}^{T} K\left(\frac{v_l - v_{t_2}}{h}\right)(v_l - v_{t_2})}{f(v_{t_2})} \right)^2 \\ &= \frac{1}{T^5} \sum_{t_1=1}^{T} \sum_{t_2=1}^{T} \frac{\zeta(v_{t_1}) \zeta(v_{t_2})}{f(v_{t_1})^2 f(v_{t_2})^2} \left(\sum_{s=1}^{T} L\left(\frac{v_s - v_{t_1}}{h}\right) \right)^2 \left(\sum_{l=1}^{T} L\left(\frac{v_l - v_{t_2}}{h}\right) \right)^2 \end{aligned}$$

$$= \frac{1}{T^5}\sum_{t_1=1}^{T}\sum_{t_2=1}^{T}\sum_{s_1=1}^{T}\sum_{s_2=1}^{T}\sum_{l_1=1}^{T}\sum_{l_2=1}^{T}\frac{\zeta(v_{t_1})\zeta(v_{t_2})}{f(v_{t_1})^2 f(v_{t_2})^2}L\left(\frac{v_{s_1}-v_{t_1}}{h}\right)L\left(\frac{v_{s_2}-v_{t_1}}{h}\right)$$

$$\times L\left(\frac{v_{l_1}-v_{t_2}}{h}\right)L\left(\frac{v_{l_2}-v_{t_2}}{h}\right). \tag{A.17}$$

Therefore, we have to prove that the expectation of

$$\frac{1}{T^5}\sum_{t_1=1}^{T}\sum_{t_2=1}^{T}\sum_{s_1=1}^{T}\sum_{s_2=1}^{T}\sum_{l_1=1}^{T}\sum_{l_2=1}^{T}\frac{\zeta(v_{t_1})\zeta(v_{t_2})}{f(v_{t_1})^2 f(v_{t_2})^2}L\left(\frac{v_{s_1}-v_{t_1}}{h}\right)$$

$$\times L\left(\frac{v_{s_2}-v_{t_1}}{h}\right)L\left(\frac{v_{l_1}-v_{t_2}}{h}\right)L\left(\frac{v_{l_2}-v_{t_2}}{h}\right) \tag{A.18}$$

is $o(1)$ and this proves

$$\mathrm{E}[\widetilde{I}_{1T}^2] = o(1), \tag{A.19}$$

which implies that $\widetilde{I}_{1T} = o_p(1)$ and therefore $I_{1T} = o_p(1)$.
For I_{2T}, it can be written as

$$I_{2T} = \frac{1}{\sqrt{T}}\sum_{t=1}^{T}\widetilde{g}(\tau_t)\zeta(v_t)\widetilde{v}_t$$

$$= \frac{1}{T^2\sqrt{T}h^2}\sum_{t=1}^{T}\zeta(v_t)\left(\frac{\sum_{s=1}^{T}K\left(\frac{v_s-v_t}{h}\right)(g(\tau_s)-g(\tau_t))}{\widehat{f}(v_t)}\right)$$

$$\left(\frac{\sum_{s=1}^{T}K\left(\frac{v_s-v_t}{h}\right)(v_s-v_t)}{\widehat{f}(v_t)}\right)$$

$$= \frac{1+o_p(1)}{T^2\sqrt{T}h^2}\sum_{t=1}^{T}\zeta(v_t)\left(\frac{\sum_{s=1}^{T}K\left(\frac{v_s-v_t}{h}\right)(g(\tau_s)-g(\tau_t))}{f(v_t)}\right)$$

$$\left(\frac{\sum_{s=1}^{T}K\left(\frac{v_s-v_t}{h}\right)(v_s-v_t)}{f(v_t)}\right)$$

$$= (1+o_p(1))\widetilde{I}_{2T}. \tag{A.20}$$

Hence, it is equivalent to prove that $\widetilde{I}_{2T}$ is $o_p(1)$. We then show that $\mathrm{E}[\widetilde{I}_{2T}^2] \to 0$ as $T \to \infty$:

$$
\begin{aligned}
\widetilde{I}_{2T}^2 &= \frac{1}{T^5h^4}\sum_{t_1=1}^{T}\sum_{t_2=1}^{T}\zeta(v_{t_1})\zeta(v_{t_2})\left(\frac{\sum_{s=1}^{T}K\left(\frac{v_s-v_{t_1}}{h}\right)(g(\tau_s)-g(\tau_{t_1}))}{f(v_{t_1})}\right)\\
&\quad\left(\frac{\sum_{s=1}^{T}K\left(\frac{v_s-v_{t_1}}{h}\right)(v_s-v_{t_1})}{f(v_{t_1})}\right)\left(\frac{\sum_{l=1}^{T}K\left(\frac{v_l-v_{t_2}}{h}\right)(g(\tau_l)-g(\tau_{t_2}))}{f(v_{t_2})}\right)\\
&\quad\left(\frac{\sum_{l=1}^{T}K\left(\frac{v_l-v_{t_2}}{h}\right)(v_l-v_{t_2})}{f(v_{t_2})}\right)\\
&= \frac{1}{T^5h^2}\sum_{t_1=1}^{T}\sum_{t_2=1}^{T}\frac{\zeta(v_{t_1})\zeta(v_{t_2})}{f(v_{t_1})^2f(v_{t_2})^2}\left(\sum_{s_1=1}^{T}K\left(\frac{v_{s_1}-v_{t_1}}{h}\right)(g(\tau_{s_1})-g(\tau_{t_1}))\right)\\
&\quad\left(\sum_{s_2=1}^{T}L\left(\frac{v_{s_2}-v_{t_1}}{h}\right)\right)\left(\sum_{l_1=1}^{T}K\left(\frac{v_{l_1}-v_{t_2}}{h}\right)(g(\tau_s)-g(\tau_{t_2}))\right)\\
&\quad\left(\sum_{l_2=1}^{T}L\left(\frac{v_{l_2}-v_{t_2}}{h}\right)\right)\\
&= \frac{1}{T^5h^2}\sum_{t_1=1}^{T}\sum_{t_2=1}^{T}\sum_{s_1=1}^{T}\sum_{s_2=1}^{T}\sum_{l_1=1}^{T}\sum_{l_2=1}^{T}\frac{\zeta(v_{t_1})\zeta(v_{t_2})}{f(v_{t_1})^2f(v_{t_2})^2}(g(\tau_{s_1})-g(\tau_{t_1}))\\
&\quad\times(g(\tau_{l_1})-g(\tau_{t_2}))K\left(\frac{v_{s_1}-v_{t_1}}{h}\right)L\left(\frac{v_{s_2}-v_{t_1}}{h}\right)K\left(\frac{v_{l_1}-v_{t_2}}{h}\right)\\
&\quad L\left(\frac{v_{l_2}-v_{t_2}}{h}\right).
\end{aligned}
\tag{A.21}
$$

Therefore, we need to show that the expectation of

$$
\begin{aligned}
&\frac{1}{T^5h^2}\sum_{t_1=1}^{T}\sum_{t_2=1}^{T}\sum_{s_1=1}^{T}\sum_{s_2=1}^{T}\sum_{l_1=1}^{T}\sum_{l_2=1}^{T}\frac{\zeta(v_{t_1})\zeta(v_{t_2})}{f(v_{t_1})^2f(v_{t_2})^2}\\
&\quad\times(g(\tau_{s_1})-g(\tau_{t_1}))(g(\tau_{l_1})-g(\tau_{t_2}))\\
&\quad K\left(\frac{v_{s_1}-v_{t_1}}{h}\right)L\left(\frac{v_{s_2}-v_{t_1}}{h}\right)K\left(\frac{v_{l_1}-v_{t_2}}{h}\right)L\left(\frac{v_{l_2}-v_{t_2}}{h}\right)
\end{aligned}
\tag{A.22}
$$

is $o(1)$, and this implies

$$\mathrm{E}[\widetilde{I}_{2T}^2] = o(1), \tag{A.23}$$

which further implies $I_{2T} = o_p(1)$.

Proof of Lemma 3.8.3. As $\breve{x}_t = \breve{g}(\tau_t) + \breve{v}_t$, we have

$$\frac{1}{\sqrt{T}}\sum_{t=1}^{T}\breve{x}_t\bar{u}_t = \frac{1}{\sqrt{T}}\sum_{t=1}^{T}\breve{g}(\tau_t)\bar{u}_t + \frac{1}{\sqrt{T}}\sum_{t=1}^{T}\breve{v}_t\bar{u}_t \triangleq G_1(T) + G_2(T), \tag{A.24}$$

hence, we need to show that both $G_1(T)$ and $G_2(T)$ are $o_p(1)$. Note that by previous definition,

$$\begin{aligned}
G_1(T) &= \frac{1}{\sqrt{T}}\sum_{t=1}^{T}\breve{g}(\tau_t)\bar{u}_t = \frac{1}{\sqrt{T}}\sum_{t=1}^{T}\left(g(\tau_t) - \sum_{s=1}^{T} w_{Ts}(t)g(\tau_s)\right)\left(\sum_{l=1}^{T} w_{Tl}(t)u_l\right)\\
&= \frac{1}{\sqrt{T}}\sum_{t=1}^{T}\left(\frac{1}{Th}\sum_{s=1}^{T}\frac{K\left(\frac{v_s-v_t}{h}\right)(g(\tau_t)-g(\tau_s))}{\widehat{f}(v_t)}\right)\left(\frac{1}{Th}\sum_{l=1}^{T}\frac{K\left(\frac{v_l-v_t}{h}\right)u_l}{\widehat{f}(v_t)}\right)\\
&= \frac{1+o_p(1)}{T^2\sqrt{T}h^2}\sum_{t=1}^{T}\sum_{s=1}^{T}\sum_{l=1}^{T}\frac{K\left(\frac{v_s-v_t}{h}\right)(g(\tau_t)-g(\tau_s))K\left(\frac{v_l-v_t}{h}\right)u_l}{f(v_t)^2}\\
&\triangleq (1+o_p(1))\widetilde{G}_1(T).
\end{aligned} \tag{A.25}$$

Therefore, it suffices to show that $\widetilde{G}_1(T)$ is $o_p(1)$. Note that

$$\begin{aligned}
\widetilde{G}_1(T)^2 &= \left(\frac{1}{T^2\sqrt{T}h^2}\sum_{t=1}^{T}\sum_{s=1}^{T}\sum_{l=1}^{T}\frac{K\left(\frac{v_s-v_t}{h}\right)(g(\tau_t)-g(\tau_s))K\left(\frac{v_l-v_t}{h}\right)u_l}{f(v_t)^2}\right)^2\\
&= \frac{1}{T^5h^4}\sum_{t_1=1}^{T}\sum_{t_2=1}^{T}\sum_{s_1=1}^{T}\sum_{s_2=1}^{T}\sum_{l_1=1}^{T}\sum_{l_2=1}^{T}\\
&\qquad \frac{K\left(\frac{v_{s_1}-v_{t_1}}{h}\right)K\left(\frac{v_{s_2}-v_{t_2}}{h}\right)K\left(\frac{v_{l_1}-v_{t_1}}{h}\right)K\left(\frac{v_{l_2}-v_{t_2}}{h}\right)}{f(v_{t_1})^2 f(v_{t_2})^2}\\
&\qquad \times u_{l_1}u_{l_2}(g(\tau_{t_1})-g(\tau_{s_1}))(g(\tau_{t_2})-g(\tau_{s_2})).
\end{aligned} \tag{A.26}$$

Therefore, it is tedious but not difficult to show that $\mathrm{E}[\widetilde{G}_1(T)^2] = o(1)$, which implies that $G_1(T) = o_p(1)$. While for $G_2(T)$,

$$
\begin{aligned}
G_2(T) =& \frac{1}{\sqrt{T}} \sum_{t=1}^{T} \widetilde{v}_t \bar{u}_t = \frac{1}{\sqrt{T}} \sum_{t=1}^{T} \left(v_t - \sum_{s=1}^{T} w_{Ts}(t) v_s \right) \left(\sum_{l=1}^{T} w_{nl}(t) u_l \right) \\
=& \frac{1}{\sqrt{T}} \sum_{t=1}^{T} \left(\frac{1}{Th} \sum_{s=1}^{T} \frac{K\left(\frac{v_s - v_t}{h}\right)(v_t - v_s)}{\widehat{f}(v_t)} \right) \left(\frac{1}{Th} \sum_{l=1}^{T} \frac{K\left(\frac{v_l - v_t}{h}\right) u_l}{\widehat{f}(v_t)} \right) \\
=& \frac{1 + o_p(1)}{n^2 \sqrt{T} h^2} \sum_{t=1}^{T} \sum_{s=1}^{T} \sum_{l=1}^{T} \frac{K\left(\frac{v_s - v_t}{h}\right)(v_t - v_s) K\left(\frac{v_l - v_t}{h}\right) u_l}{f(v_t)^2} \\
\triangleq& (1 + o_p(1)) \widetilde{G}_2(T).
\end{aligned} \tag{A.27}
$$

Thus, it is sufficient to show $\widetilde{G}_2(T) = o_p(1)$. Let $L(u) = uK(u)$, and note that

$$
\begin{aligned}
\widetilde{G}_2(T)^2 =& \left(\frac{1}{T^2 \sqrt{T} h} \sum_{t=1}^{T} \sum_{s=1}^{T} \sum_{l=1}^{T} \frac{L\left(\frac{v_s - v_t}{h}\right) K\left(\frac{v_l - v_t}{h}\right) u_l}{f(v_t)^2} \right)^2 \\
=& \frac{1}{T^5 h^2} \sum_{t_1=1}^{T} \sum_{t_2=1}^{T} \sum_{s_1=1}^{T} \sum_{s_2=1}^{T} \sum_{l_1=1}^{T} \sum_{l_2=1}^{T} \\
& \frac{L\left(\frac{v_{s_1} - v_{t_1}}{h}\right) K\left(\frac{v_{s_2} - v_{t_2}}{h}\right) L\left(\frac{v_{l_1} - v_{t_1}}{h}\right) K\left(\frac{v_{l_2} - v_{t_2}}{h}\right) u_{l_1} u_{l_2}}{f(v_{t_1})^2 f(v_{t_2})^2}.
\end{aligned} \tag{A.28}
$$

Similar to the previous proof, we are able to show that $\mathrm{E}\left[\widetilde{G}_2(T)^2\right] = o(1)$, which implies that $G_2(T) = o_p(1)$. Therefore,

$$
\frac{1}{\sqrt{T}} \sum_{t=1}^{T} \check{x}_t \bar{u}_t = G_1(T) + G_2(T) = o_p(1). \tag{A.29}
$$

Proof of Lemma 3.8.4. Without loss of generality (proofs for $k > 1$ the multivariate case can be carried out at the cost of more tedious derivations), we assume $k = 1$. Therefore, our objective is to prove that

$$\mathfrak{D}_1(T) = \frac{1}{T}\sum_{t=1}^{T}(\widehat{x}_t - \check{x}_t)^2 = o_p(1). \tag{A.30}$$

Substitute $\widehat{x}_t$ and $\check{x}_t$ in $\mathfrak{D}_1(T)$,

$$\begin{aligned}
\mathfrak{D}_1(T) &= \frac{1}{T}\sum_{t=1}^{T}(\widehat{x}_t - \check{x}_t)^2 \\
&= \frac{1}{T}\sum_{t=1}^{T}\left(x_t - \sum_{s=1}^{T}\bar{w}_{Ts}(t)x_s - x_t + \sum_{s=1}^{T} w_{Ts}(t)x_s\right)^2 \\
&= \frac{1}{T}\sum_{t=1}^{T}\left(\sum_{s=1}^{T}\left(w_{Ts}(t) - \bar{w}_{Ts}(t)\right)x_s\right)^2 \\
&= \frac{1}{T}\sum_{t=1}^{T}\sum_{s=1}^{T}\left(w_{Ts}(t) - \bar{w}_{Ts}(t)\right)^2 x_s^2 \\
&\quad + \frac{1}{T}\sum_{t=1}^{T}\sum_{s=1}^{T}\sum_{\substack{r=1 \\ r\neq s}}^{T}\left(w_{Ts}(t) - \bar{w}_{Ts}(t)\right)\left(w_{Tr}(t) - \bar{w}_{Tr}(t)\right)x_s x_r \\
&= \mathfrak{D}_{11}(T) + \mathfrak{D}_{12}(T).
\end{aligned} \tag{A.31}$$

Let $\Pi_{s,t} = w_{Ts}(t) - \bar{w}_{Ts}(t)$. We define several useful notations as follows:

$$\begin{aligned}
\Xi_1(r,t) &= \frac{K\left(\frac{v_r - v_t}{h}\right)}{f(v_t)}(g(\tau_t) - g(\tau_r)), \\
\Xi_2(r,t) &= \frac{K\left(\frac{v_r - v_t}{h}\right)}{f(v_t)}(v_t - v_r), \\
\Xi_3(l,t) &= K_2\left(\frac{\tau_l - \tau_t}{b}\right)(g(\tau_l) - g(\tau_t)), \\
\Xi_4(l,t) &= K_2\left(\frac{\tau_l - \tau_t}{b}\right)v_t, \\
\Xi_5(p,t,l) &= \Xi_3(l,p) + \Xi_4(l,p) - \Xi_3(l,t) - \Xi_4(l,t).
\end{aligned}$$

Note that

$$\begin{aligned}
\Pi_{s,t} =& \frac{K\left(\frac{v_s-v_t}{h}\right)}{Th\widehat{f}(v_t)} - \frac{K\left(\frac{\widehat{v}_s-\widehat{v}_t}{h}\right)}{Th\widehat{f}(\widehat{v}_t)} = \frac{K\left(\frac{v_s-v_t}{h}\right)(\widehat{f}(v_t)-\widehat{f}(\widehat{v}_t))}{Th\widehat{f}(v_t)\widehat{f}(\widehat{v}_t)} \\
&+ \frac{\left(K\left(\frac{v_t-v_s}{h}\right) - K\left(\frac{\widehat{v}_t-\widehat{v}_s}{h}\right)\right)}{Th\widehat{f}(\widehat{v}_t)} \\
=& \frac{K\left(\frac{v_s-v_t}{h}\right)(\widehat{f}(v_t)-\widehat{f}(\widehat{v}_t))}{Th(f(v_t)^2+o_p(1))} + \frac{\left(K\left(\frac{v_t-v_s}{h}\right) - K\left(\frac{\widehat{v}_t-\widehat{v}_s}{h}\right)\right)}{Th(f(v_t)+o_p(1))} \\
=&(1+o_p(1))\left(\frac{K\left(\frac{v_s-v_t}{h}\right)(\widehat{f}(v_t)-\widehat{f}(\widehat{v}_t))}{Thf(v_t)^2} + \frac{\left(K\left(\frac{v_t-v_s}{h}\right) - K\left(\frac{\widehat{v}_t-\widehat{v}_s}{h}\right)\right)}{Thf(v_t)}\right).
\end{aligned} \tag{A.32}$$

By Taylor expansion, and denote $K_d(p,t) = K'\left(\frac{v_p-v_t}{h}\right)$,

$$\begin{aligned}
&K\left(\frac{v_p-v_t}{h}\right) - K\left(\frac{\widehat{v}_p-\widehat{v}_t}{h}\right) \\
=&(1+o_p(1))K'\left(\frac{v_p-v_t}{h}\right)\left(\left(\frac{v_p-v_t}{h}\right) - \left(\frac{\widehat{v}_p-\widehat{v}_t}{h}\right)\right) \\
=&\frac{(1+o_p(1))^2}{Thb}K_d(p,t)\left(\sum_{l=1}^T K_2\left(\frac{\tau_l-\tau_p}{b}\right)(g(\tau_l)-g(\tau_p)) + \sum_{l=1}^T K_2\left(\frac{\tau_l-\tau_p}{b}\right)v_l\right) \\
&- \frac{(1+o_p(1))^2}{Thb}K_d(p,t)\left(\sum_{l=1}^T K_2\left(\frac{\tau_l-\tau_t}{b}\right)(g(\tau_l)-g(\tau_t)) + \sum_{l=1}^T K_2\left(\frac{\tau_l-\tau_t}{b}\right)v_t\right) \\
\triangleq&\frac{(1+o_p(1))^2}{Thb}K_d(p,t)\sum_{l=1}^T\left(\Xi_3(l,p)+\Xi_4(l,p)-\Xi_3(l,t)-\Xi_4(l,t)\right) \\
\triangleq&\frac{(1+o_p(1))^2}{Thb}K_d(p,t)\sum_{l=1}^T\Xi_5(p,t,l),
\end{aligned} \tag{A.33}$$

where $\Xi_5(p,t,l) = \Xi_3(l,p)+\Xi_4(l,p)-\Xi_3(l,t)-\Xi_4(l,t)$. Meanwhile,

$$\begin{aligned}
\widehat{f}(v_t) - \widehat{f}(\widehat{v}_t) =& \frac{1}{Th}\sum_{p=1}^T\left(K\left(\frac{v_p-v_t}{h}\right) - K\left(\frac{\widehat{v}_p-\widehat{v}_t}{h}\right)\right) \\
=& \frac{(1+o_p(1))}{T^2h^2b}\sum_{p=1}^T K_d(p,t)\sum_{l=1}^T\Xi_5(p,t,l).
\end{aligned} \tag{A.34}$$

Therefore,

$$\begin{aligned}\Pi_{s,t} &= (1+o_p(1))\left(\frac{K\left(\frac{v_s-v_t}{h}\right)}{T^3h^3bf(v_t)^2}\sum_{p=1}^{T}K_d(p,t)\sum_{l=1}^{T}\Xi_5(p,t,l)\right)\\ &\quad+(1+o_p(1))\left(\frac{K_d(s,t)\sum_{l=1}^{T}\Xi_5(s,t,l)}{T^2h^2bf(v_t)}\right)\\ &\triangleq (1+o_p(1))(\Pi_{s,t,1}+\Pi_{s,t,2}). \end{aligned} \tag{A.35}$$

We then have

$$\begin{aligned}\mathfrak{D}_{11}(T) &= \frac{1}{T}\sum_{t=1}^{T}\sum_{s=1}^{T}\left(w_{Ts}(t)-\bar{w}_{Ts}(t)\right)^2 x_s^2\\ &= \frac{1}{T}\sum_{t=1}^{T}\sum_{s=1}^{T}\Big((1+o_p(1))(\Pi_{s,t,1}+\Pi_{s,t,2})\Big)^2 x_s^2\\ &= \frac{(1+o_p(1))}{T}\sum_{t=1}^{T}\sum_{s=1}^{T}(\Pi_{s,t,1}+\Pi_{s,t,2})^2x_s^2 \triangleq (1+o_p(1))\widetilde{\mathfrak{D}}_{11}. \end{aligned} \tag{A.36}$$

Also note that

$$\begin{aligned}\widetilde{\mathfrak{D}}_{11} &= \frac{1}{T}\sum_{t=1}^{T}\sum_{s=1}^{T}(\Pi_{s,t,1}+\Pi_{s,t,2})^2x_s^2\\ &\leq \frac{2}{n}\sum_{t=1}^{T}\sum_{s=1}^{T}\Pi_{s,t,1}^2x_s^2+\frac{2}{n}\sum_{t=1}^{T}\sum_{s=1}^{T}\Pi_{s,t,2}^2x_s^2 \triangleq 2\widetilde{\mathfrak{D}}_{11,1}+2\widetilde{\mathfrak{D}}_{11,2}. \end{aligned} \tag{A.37}$$

Hence, we need to show that $\widetilde{\mathfrak{D}}_{11,1} = o_p(1)$ and $\widetilde{\mathfrak{D}}_{11,2} = o_p(1)$. As the all the terms in $\widetilde{\mathfrak{D}}_{11,i}$ are positive, we only need to show that $\mathrm{E}[\widetilde{\mathfrak{D}}_{11,i}] \to 0$ as $T \to \infty$ for $i = 1, 2$. We omit the proofs here as they belong to special and simpler cases of our subsequent proofs for

$\mathfrak{D}_{12}(T)$, where

$$\begin{aligned}
\mathfrak{D}_{12}(T) &= \frac{1}{T}\sum_{t=1}^{T}\sum_{s=1}^{T}\sum_{\substack{r=1\\ r\neq s}}^{T}\left(w_{Ts}(t)-\bar{w}_{ns}(t)\right)\left(w_{nr}(t)-\bar{w}_{nr}(t)\right)x_s x_r \\
&= \frac{(1+o_p(1))}{T}\sum_{t=1}^{T}\sum_{s=1}^{T}\sum_{\substack{r=1\\ r\neq s}}^{T}\left(\Pi_{s,t,1}+\Pi_{s,t,2}\right)\left(\Pi_{r,t,1}+\Pi_{r,t,2}\right)x_s x_r \\
&= (1+o_p(1))\left(\frac{1}{T}\sum_{t=1}^{T}\sum_{s=1}^{T}\sum_{\substack{r=1\\ r\neq s}}^{T}\Pi_{s,t,1}\Pi_{r,t,1}x_s x_r\right. \\
&\quad + \frac{1}{T}\sum_{t=1}^{T}\sum_{s=1}^{T}\sum_{\substack{r=1\\ r\neq s}}^{T}\Pi_{s,t,1}\Pi_{r,t,2}x_s x_r + \frac{1}{T}\sum_{t=1}^{T}\sum_{s=1}^{T}\sum_{\substack{r=1\\ r\neq s}}^{T}\Pi_{s,t,2}\Pi_{r,t,1}x_s x_r \\
&\quad \left. + \frac{1}{T}\sum_{t=1}^{T}\sum_{s=1}^{T}\sum_{\substack{r=1\\ r\neq s}}^{T}\Pi_{s,t,2}\Pi_{r,t,2}x_s x_r\right) \\
&\triangleq (1+o_p(1))\left(\widetilde{\mathfrak{D}}_{12,1}(T)+\widetilde{\mathfrak{D}}_{12,2}(T)+\widetilde{\mathfrak{D}}_{12,3}(T)+\widetilde{\mathfrak{D}}_{12,4}(T)\right).
\end{aligned} \tag{A.38}$$

Therefore, after showing that $\widetilde{\mathfrak{D}}_{12,i}(T) = o_p(1)$ for $i = 1, 2, 3, 4$, we are able to prove that $\mathfrak{D}_{11}(T)$ and $D_{12}(T)$ are $o_p(1)$. Hence, $\mathfrak{D}_1 = o_p(1)$.

Proof of Lemma 3.8.5.
We first decompose the equation as

$$\begin{aligned}
\frac{1}{\sqrt{T}}\left(\widehat{X}'\widehat{e}-\check{X}'\check{e}\right) &= \frac{1}{\sqrt{T}}\left(\widehat{X}'\widehat{e}-\check{X}'\widehat{e}+\check{X}'\widehat{e}-\check{X}'\check{e}\right) \\
&= \frac{1}{\sqrt{T}}\left(\widehat{X}'\widehat{e}-\check{X}'\widehat{e}\right)+\frac{1}{\sqrt{T}}\left(\check{X}'\widehat{e}-\check{X}'\check{e}\right) \\
&= \frac{1}{\sqrt{T}}\left(\widehat{X}-\check{X}\right)'\widehat{e}+\frac{1}{\sqrt{T}}\check{X}'\left(\widehat{e}-\check{e}\right)
\end{aligned}$$

$$
\begin{aligned}
&= \frac{1}{\sqrt{T}}\left(\widehat{X}-\check{X}\right)'(\widehat{e}-\check{e}) \\
&\quad + \frac{1}{\sqrt{T}}\left(\widehat{X}-\check{X}\right)'\check{e} + \frac{1}{\sqrt{T}}\check{X}'(\widehat{e}-\check{e}) \\
&= P_1(T)+P_2(T)+P_3(T), \qquad \text{(A.39)}
\end{aligned}
$$

where $P_1(T) = \left(\widehat{X}-\check{X}\right)'(\widehat{e}-\check{e})/\sqrt{T}$, $P_2(T) = \left(\widehat{X}-\check{X}\right)'\check{e}/\sqrt{T}$, $P_3(T) = \check{X}'(\widehat{e}-\check{e})/\sqrt{T}$. Therefore, it suffices to show that $P_i(T) = o_p(1)$ as $T \to \infty$ for $i = 1, 2, 3$. Note that $\widehat{e} = \widehat{\lambda}(V)+\widehat{U}$, $\check{e} = \check{\lambda}(V)+\check{U}$, and write $\lambda_t = \lambda(v_t)$. We have

$$
\begin{aligned}
P_1(T) &= \frac{1}{\sqrt{T}}\left(\widehat{X}-\check{X}\right)'(\widehat{e}-\check{e}) = \frac{1}{\sqrt{T}}\sum_{t=1}^{T}(\widehat{x}_t-\check{x}_t)(\widehat{e}_t-\check{e}_t) \\
&= \frac{1}{\sqrt{T}}\sum_{t=1}^{T}(\widehat{x}_t-\check{x}_t)(\widehat{\lambda}_t-\check{\lambda}_t) + \frac{1}{\sqrt{T}}\sum_{t=1}^{T}(\widehat{x}_t-\check{x}_t)(\widehat{u}_t-\check{U}_t) \\
&= P_{11}(T)+P_{12}(T), \qquad \text{(A.40)}
\end{aligned}
$$

$$
\begin{aligned}
P_2(T) &= \frac{1}{\sqrt{T}}\left(\widehat{X}-\check{X}\right)'\check{e} = \frac{1}{\sqrt{T}}\sum_{t=1}^{T}(\widehat{x}_t-\check{x}_t)\check{e}_t \\
&= \frac{1}{\sqrt{T}}\sum_{t=1}^{T}(\widehat{x}_t-\check{x}_t)\check{\lambda}_t + \frac{1}{\sqrt{T}}\sum_{t=1}^{T}(\widehat{x}_t-\check{x}_t)\check{U}_t \\
&= P_{21}(T)+P_{22}(T), \qquad \text{(A.41)}
\end{aligned}
$$

$$
\begin{aligned}
P_3(T) &= \frac{1}{\sqrt{T}}\check{X}'(\widehat{e}-\check{e}) = \frac{1}{\sqrt{T}}\sum_{t=1}^{T}\check{x}_t(\widehat{e}_t-\check{e}_t) \\
&= \frac{1}{\sqrt{T}}\sum_{t=1}^{T}\check{x}_t(\widehat{\lambda}_t-\check{\lambda}_t) + \frac{1}{\sqrt{T}}\sum_{t=1}^{T}\check{x}_t(\widehat{u}_t-\check{U}_t) \\
&= P_{31}(T)+P_{32}(T). \qquad \text{(A.42)}
\end{aligned}
$$

Hence, once we have shown that $P_{ij}(T) = o_p(1)$ for $i = 1, 2, 3, j = 1, 2$, we are able to complete the proof.

A.2 Proofs of the Lemmas in Chapter 4

Proof of Lemma 4.8.1. Note that

$$\begin{aligned}
\frac{1}{T^{\frac{d_i}{2}}}\sum_{t=1}^{T} g_i(t)(\widetilde{\epsilon}_{t-1}-\widetilde{\epsilon}_t) &= \frac{1}{\sqrt{T}}\sum_{t=1}^{T}\frac{g_i(t)}{T^{\frac{d_i-1}{2}}}(\widetilde{\epsilon}_{t-1}-\widetilde{\epsilon}_t) \\
&= \frac{1}{\sqrt{T}}\sum_{t=1}^{T}\widetilde{g}_i(\tau_t)(\widetilde{\epsilon}_{t-1}-\widetilde{\epsilon}_t) = \frac{1}{\sqrt{T}}\sum_{t=1}^{T}\Big(\widetilde{g}_i(\tau_t)\widetilde{\epsilon}_{t-1}-\widetilde{g}_i(\tau_t)\widetilde{\epsilon}_t\Big) \\
&= \frac{1}{\sqrt{T}}\sum_{t=1}^{T}\Big(\widetilde{g}_i(\tau_t)\widetilde{\epsilon}_{t-1}-\widetilde{g}_i(\tau_{t-1})\widetilde{\epsilon}_{t-1}+\widetilde{g}_i(\tau_{t-1})\widetilde{\epsilon}_{t-1}-\widetilde{g}_i(\tau_t)\widetilde{\epsilon}_t\Big) \\
&= \frac{1}{\sqrt{T}}\sum_{t=1}^{T}\Big(\widetilde{g}_i(\tau_t)\widetilde{\epsilon}_{t-1}-\widetilde{g}_i(\tau_{t-1})\widetilde{\epsilon}_{t-1}\Big) \\
&\quad + \frac{1}{\sqrt{T}}\sum_{t=1}^{T}\Big(\widetilde{g}_i(\tau_{t-1})\widetilde{\epsilon}_{t-1}-\widetilde{g}_i(\tau_t)\widetilde{\epsilon}_t\Big) \\
&= \frac{1}{\sqrt{T}}\sum_{t=1}^{T}\Big(\widetilde{g}_i(\tau_t)-\widetilde{g}_i(\tau_{t-1})\Big)\widetilde{\epsilon}_{t-1}+\frac{1}{\sqrt{T}}\Big(\widetilde{g}_i(\tau_0)\widetilde{\epsilon}_0-\widetilde{g}_i(\tau_n)\widetilde{\epsilon}_T\Big),
\end{aligned} \tag{A.43}$$

where the second term is $o_p(1/\sqrt{T})$. For the first term, by Taylor expansion,

$$\zeta_3(T) = \frac{1}{\sqrt{T}}\sum_{t=1}^{T}\Big(\widetilde{g}_i(\tau_t)-\widetilde{g}_i(\tau_{t-1})\Big)\widetilde{\epsilon}_{t-1} = \frac{1}{T\sqrt{T}}\sum_{t=1}^{T}\widetilde{g}_i'(\tau_{t-1})\widetilde{\epsilon}_{t-1}. \tag{A.44}$$

It is easy to show that $\mathrm{E}[\zeta_3(T)] = 0$ and $\mathrm{E}[\zeta_3(T)^2] = O(T^{-2})$. Therefore, as $T \to \infty$, we have

$$\frac{1}{\sqrt{T}}\sum_{t=1}^{T}\widetilde{g}_i(\tau_t)(\widetilde{\epsilon}_{t-1}-\widetilde{\epsilon}_t) \longrightarrow_P 0.$$

We then complete the proof.

Proof of Lemma 4.8.2. We need to show that as $T \to \infty$,

$$\frac{1}{T^{d_i/2}} \widetilde{\epsilon\eta}_{i,0}^{f} \longrightarrow_P 0, \tag{A.45}$$

$$\frac{1}{T^{d_i/2}} \widetilde{\epsilon\eta}_{i,n}^{f} \longrightarrow_P 0, \tag{A.46}$$

where $\widetilde{\epsilon\eta}_{i,t}^{f} = \widetilde{f}_{i,0}(L)(\epsilon_t\eta_{it} - \theta_i) = \sum_{s=0}^{\infty} \widetilde{f}_{i,0s}(\epsilon_{t-s}\eta_{i,t-s} - \theta_i)$ and $\widetilde{f}_{i,0s} = \sum_{k=s+1}^{\infty} \phi_k \psi_{k,i}$. It is obvious that

$$\begin{aligned} \mathrm{E}\left[\frac{1}{T^{d_i/2}} \widetilde{\epsilon\eta}_{i,n}^{f}\right] &= \frac{1}{T^{d_i/2}} \mathrm{E}\left[\sum_{s=0}^{\infty} \widetilde{f}_{i,0s}(\epsilon_{t-s}\eta_{i,t-s} - \theta_i)\right] \\ &= \frac{1}{T^{d_i/2}} \sum_{s=0}^{\infty} \widetilde{f}_{i,0s}(\mathrm{E}\left[\epsilon_{t-s}\eta_{i,t-s}\right] - \theta_i) = 0. \end{aligned} \tag{A.47}$$

Meanwhile,

$$\begin{aligned} &\mathrm{E}\left[\left(\frac{1}{T^{d_i/2}} \widetilde{\epsilon\eta}_{i,n}^{f}\right)^2\right] \\ &= \frac{1}{T^{d_i}} \sum_{s=0}^{\infty} \widetilde{f}_{i,0s}^{2} \mathrm{E}[(\epsilon_{t-s}\eta_{i,t-s} - \theta_i)^2] \\ &\quad + \frac{2}{T^{d_i}} \sum_{s_1=0}^{\infty} \sum_{s_2=s_1+1}^{\infty} \widetilde{f}_{i,0s_1} \widetilde{f}_{i,0s_2} \mathrm{E}[(\epsilon_{t-s_1}\eta_{i,t-s_1} - \theta_i)(\epsilon_{t-s_2}\eta_{i,t-s_2} - \theta_i)] \\ &= \frac{\widetilde{\delta}_{22} - \theta_i^2}{T^{d_i}} \sum_{s=0}^{\infty} \widetilde{f}_{i,0s}^{2} = O(T^{-d_i}), \end{aligned} \tag{A.48}$$

given that $\sum_{s=0}^{\infty} \widetilde{f}_{i,0s}^{2} < \infty$. Therefore, as $T \to \infty$, (A.45) holds. Similarly, we can prove (A.46) given the same condition. Thus, we complete the proof.

Proof of Lemmas 4.8.3 and 4.8.4. In these two lemmas, we define

$$B_{i,t}^{f} = \sum_{q=1}^{\infty} \widetilde{f}_{i,q}(L)\epsilon_t\eta_{i,t-q}, \tag{A.49}$$

$$B_{i,t}^{m} = \sum_{q=1}^{\infty} \widetilde{m}_{i,q}(L)\epsilon_{t-q}\eta_{it}, \tag{A.50}$$

where $\widetilde{f}_{i,q}(L) = \sum_{s=0}^{\infty} \widetilde{f}_{i,qs} L^s$, $\widetilde{f}_{i,qs} = \sum_{p=s+1}^{\infty} \phi_p \psi_{p+q,i}$ and $\widetilde{m}_{i,q}(L) = \sum_{s=0}^{\infty} \widetilde{m}_{i,qs} L^s$, $\widetilde{m}_{i,qs} = \sum_{p=s+1}^{\infty} \phi_{p+q} \psi_{p,i}$. Therefore, for any t,

$$\begin{aligned}
\mathrm{E}\left[\frac{1}{T^{d_i/2}} B_{i,t}^f\right] &= \mathrm{E}\left[\frac{1}{T^{d_i/2}} \sum_{q=1}^{\infty} \sum_{s=0}^{\infty} f_{i,qs} \epsilon_{t-s} \eta_{i,t-q-s}\right] \\
&= \frac{1}{T^{d_i/2}} \sum_{q=1}^{\infty} \sum_{s=0}^{\infty} f_{i,ks} \mathrm{E}[\epsilon_{t-s} \eta_{i,t-q-s}] = 0 \qquad \text{(A.51)}
\end{aligned}$$

and

$$\begin{aligned}
\mathrm{E}\left[\frac{1}{T^{d_i/2}} B_{i,t}^m\right] &= \mathrm{E}\left[\frac{1}{T^{d_i/2}} \sum_{q=1}^{\infty} \sum_{s=0}^{\infty} m_{i,qs} \epsilon_{t-q-s} \eta_{i,t-s}\right] \\
&= \frac{1}{T^{d_i/2}} \sum_{q=1}^{\infty} \sum_{s=0}^{\infty} m_{i,qs} \mathrm{E}[\epsilon_{t-q-s} \eta_{i,t-s}] = 0. \qquad \text{(A.52)}
\end{aligned}$$

For the second moment,

$$\mathrm{E}\left[\left(\frac{1}{T^{d_i/2}} B_{i,t}^f\right)^2\right] = \frac{1}{T^{d_i}} \sum_{q=1}^{\infty} \sum_{s=0}^{\infty} f_{i,qs}^2 \mathrm{E}[\epsilon_{t-s}^2 \eta_{i,t-q-s}^2] = O(T^{-d_i}), \qquad \text{(A.53)}$$

given that $\sum_{q=1}^{\infty} \sum_{s=0}^{\infty} \widetilde{f}_{i,qs}^2 < \infty$. Similarly, we can show that

$$\mathrm{E}\left[\left(\frac{1}{T^{d_i/2}} B_{i,t}^m\right)^2\right] = \frac{1}{T^{d_i}} \sum_{q=1}^{\infty} \sum_{s=0}^{\infty} m_{i,qs}^2 \mathrm{E}[\epsilon_{t-s-q}^2 \eta_{i,t-s}^2] = O(T^{-d_i}), \qquad \text{(A.54)}$$

given that $\sum_{q=1}^{\infty} \sum_{s=0}^{\infty} \widetilde{m}_{i,qs}^2 < \infty$. Therefore,

$$\frac{1}{T^{d_i/2}} B_{i,t}^m = o_P(1), \qquad \text{(A.55)}$$

$$\frac{1}{T^{d_i/2}} B_{i,t}^f = o_P(1), \qquad \text{(A.56)}$$

for $t = 0$ and $t = T$. This completes the proof for Lemmas 4.8.3 and 4.8.4.

Proof of Lemma 4.8.5. Denote $\boldsymbol{Z}_3 = (a' M_{Tt})^2$, then

$$\boldsymbol{Z}_3 = \sum_{i=1}^{k} \sum_{j=1}^{K} a_i a_j M_{Tt}^i M_{Tt}^j. \tag{A.57}$$

Note that

$$\begin{aligned}
&M_{Tt}^i M_{Tt}^j \\
&= T^{-d_i/2} \left(g_i(\tau_t)\Phi(1)\epsilon_t + f_{i,0}(1)\left(\epsilon_t \eta_{it} - \theta_i\right) + \sum_{q=1}^{\infty} f_{i,q}(1)\epsilon_t \eta_{i,t-q} \right. \\
&\qquad \left. + \sum_{q=1}^{\infty} m_{i,q}(1)\epsilon_{t-q}\eta_{i,t} \right) \\
&\quad \cdot T^{-d_j/2} \left(g_j(\tau_t)\Phi(1)\epsilon_t + f_{j,0}(1)\left(\epsilon_t \eta_{jt} - \theta_j\right) + \sum_{q=1}^{\infty} f_{j,q}(1)\epsilon_t \eta_{j,t-q} \right. \\
&\qquad \left. + \sum_{q=1}^{\infty} m_{j,q}(1)\epsilon_{t-q}\eta_{j,t} \right) \\
&\triangleq \left(M_{1,nt}^i + M_{2,Tt}^i + M_{3,nt}^i + M_{4,nt}^i\right)\left(M_{1,nt}^j + M_{2,Tt}^j + M_{3,nt}^j + M_{4,nt}^j\right),
\end{aligned} \tag{A.58}$$

where for $i = 1, 2, \ldots, k$,

$$M_{1,nt}^i = T^{-d_i/2} g_i(\tau_t)\Phi(1)\epsilon_t, \tag{A.59}$$

$$M_{2,Tt}^i = T^{-d_i/2} f_{i,0}(1)\left(\epsilon_t \eta_{it} - \theta_i\right), \tag{A.60}$$

$$M_{3,nt}^i = T^{-d_i/2} \sum_{q=1}^{\infty} f_{i,q}(1)\epsilon_t \eta_{i,t-q}, \tag{A.61}$$

$$M_{4,nt}^i = T^{-d_i/2} \sum_{q=1}^{\infty} m_{i,q}(1)\epsilon_{t-q}\eta_{i,t}. \tag{A.62}$$

Then

$$\sum_{t=1}^{T} \mathrm{E}\left[(a' M_{Tt})^2 \middle| F_{t-1}\right] = \sum_{t=1}^{T}\sum_{i=1}^{k}\sum_{j=1}^{k}\sum_{r_1=1}^{4}\sum_{r_2=1}^{4} \mathrm{E}\left[a_i a_j M^i_{r_1,Tt} M^j_{r_2,nt} \middle| F_{t-1}\right]$$

$$= \sum_{i=1}^{k}\sum_{j=1}^{k} a_i a_j \sum_{r_1=1}^{4}\sum_{r_2=1}^{4} \frac{1}{T^{d_{ij}}} \sum_{t=1}^{T} \mathrm{E}\left[M^i_{r_1,Tt} M^j_{r_2,nt} \middle| F_{t-1}\right]$$

$$= \sum_{i=1}^{k}\sum_{j=1}^{k} a_i a_j \sum_{r_1=1}^{4}\sum_{r_2=1}^{4} \boldsymbol{Z}_4(i,j,r_1,r_2), \tag{A.63}$$

where $\boldsymbol{Z}_4(i,j,r_1,r_2) \triangleq T^{-d_{ij}} \sum_{t=1}^{T} \mathrm{E}\left[M^i_{r_1,Tt} M^j_{r_2,nt} \middle| F_{t-1}\right]$. For given $i,j = 1,2,\ldots,k$, we analyze the terms in (A.63) one by one with respect to $r_1, r_2 = 1,2,3,4$:

(1) When $r_1 = 1, r_2 = 1$, as $T \to \infty$, we have

$$\sum_{t=1}^{T} \mathrm{E}\left[M^i_{1,Tt} M^j_{1,nt} \middle| F_{t-1}\right]$$

$$= \frac{1}{T^{d_{ij}}} \sum_{t=1}^{T} \mathrm{E}[g_i(\tau_t) g_j(\tau_t) \Phi(1)^2 \epsilon_t^2 | F_{t-1}]$$

$$= \frac{\sigma_1^2 \Phi(1)^2}{T^{d_{ij}}} \sum_{t=1}^{T} g_i(\tau_t) g_j(\tau_t) \longrightarrow \sigma_1^2 \Phi(1)^2 \boldsymbol{Q}_{ij}. \tag{A.64}$$

(2) When $r_1 = 2, r_2 = 2$, as $T \to \infty$, we have

$$\sum_{t=1}^{T} \mathrm{E}\left[M^i_{2,Tt} M^j_{2,Tt} \middle| F_{t-1}\right]$$

$$= f_{i,0}(1) f_{j,0}(1) \frac{1}{T^{d_{ij}}} \sum_{t=1}^{T} \mathrm{E}[(\epsilon_t \eta_{it} - \theta_i)(\epsilon_t \eta_{jt} - \theta_j)], \tag{A.65}$$

when $d_{ij} = 1$,

$$\sum_{t=1}^{T} \mathrm{E}\left[M^i_{2,Tt} M^j_{2,Tt} \middle| F_{t-1}\right] = f_{i,0}(1) f_{j,0}(1) (\mathrm{E}[\epsilon_t^2 \eta_{it} \eta_{jt}] - \theta_i \theta_j)$$

$$= f_{i,0}(1) f_{j,0}(1) (\delta_{2ij} - \theta_i \theta_j), \tag{A.66}$$

where $\delta_{2ij} = \mathrm{E}[\epsilon_t^2 \eta_{it} \eta_{jt}]$.

When $d_{ij} > 1$,

$$\sum_{t=1}^{T} \mathrm{E}\left[M_{2,Tt}^{i} M_{2,Tt}^{j} \middle| F_{t-1}\right] = f_{i,0}(1) f_{j,0}(1) \frac{\mathrm{E}[\epsilon_t^2 \eta_{it} \eta_{jt}] - \theta_i \theta_j}{T^{d_{ij}-1}} \longrightarrow 0. \tag{A.67}$$

(3) When $r_1 = 1, r_2 = 2$ or $r_1 = 2, r_2 = 1$, as $T \to \infty$, we have

$$\begin{aligned}
&\sum_{t=1}^{T} \mathrm{E}\left[M_{1,Tt}^{i} M_{2,Tt}^{j} \middle| F_{t-1}\right] \\
&= \frac{1}{T^{d_{ij}}} \sum_{t=1}^{T} \mathrm{E}[g_i(\tau_t) \Phi(1) \epsilon_t f_{j,0}(1) \left(\epsilon_t \eta_{jt} - \theta_j\right) | F_{t-1}] \\
&= \frac{\Phi(1) f_{j,0}(1)}{T^{d_{ij}}} \sum_{t=1}^{T} g_i(\tau_t) \mathrm{E}[\epsilon_t \left(\epsilon_t \eta_{jt} - \theta_j\right)] \\
&= \Phi(1) f_{j,0}(1) \delta_{2j} \frac{1}{T^{d_{ij}}} \sum_{t=1}^{T} g_i(\tau_t) \\
&= \Phi(1) f_{j,0}(1) \delta_{2j} \frac{1}{T^{(d_j-1)/2}} \frac{1}{T} \sum_{t=1}^{T} \frac{g_i(\tau_t)}{T^{(d_i-1)/2}},
\end{aligned} \tag{A.68}$$

where $\delta_{2j} = \mathrm{E}[\epsilon_t^2 \eta_{jt}]$. When $d_j = 1$,

$$\sum_{t=1}^{T} \mathrm{E}\left[M_{1,Tt}^{i} M_{2,Tt}^{j} \middle| F_{t-1}\right] \longrightarrow \Phi(1) f_{j,0}(1) \delta_{2j} \bar{g}_i, \tag{A.69}$$

in which $\bar{g}_i = \int_0^1 g_i^N(\tau) d\tau$.
When $d_j > 1$,

$$\sum_{t=1}^{T} \sum_{i=1}^{K} \sum_{j=1}^{K} \mathrm{E}\left[a_i a_j M_{1,Tt}^{i} M_{2,Tt}^{j} \middle| F_{t-1}\right] \longrightarrow 0. \tag{A.70}$$

Similar result holds for $r_1 = 2, r_2 = 1$.

(4) When $r_1 = r_2 = 3$, as $T \to \infty$, we have

$$\begin{aligned}
&\sum_{t=1}^{T} \mathrm{E}\left[M_{3,Tt}^{i} M_{3,nt}^{j} \middle| F_{t-1}\right] \\
&= \frac{1}{T^{d_{ij}}} \sum_{t=1}^{T} \mathrm{E}\left[\sum_{q_1=1}^{\infty} f_{i,q_1}(1)\epsilon_t \eta_{i,t-q_1} \sum_{q_2=1}^{\infty} f_{j,q_2}(1)\epsilon_t \eta_{j,t-q_2} \middle| F_{t-1}\right] \\
&= \frac{\sigma_1^2}{T^{d_{ij}}} \sum_{t=1}^{T} \sum_{q_1=1}^{\infty} \sum_{q_2=1}^{\infty} f_{i,q_1}(1) f_{j,q_2}(1) \eta_{i,t-q_1} \eta_{j,t-q_2} \\
&= \frac{\sigma_1^2}{T^{d_{ij}}} \sum_{t=1}^{T} \sum_{q_1=1}^{\infty} f_{i,q_1}(1) f_{j,q_1}(1) \eta_{i,t-q_1} \eta_{j,t-q_1} \\
&\quad + \frac{\sigma_1^2}{T^{d_{ij}}} \sum_{t=1}^{T} \sum_{q_1=1}^{\infty} \sum_{\substack{q_2=1 \\ q_2 \neq q_1}}^{\infty} f_{i,q_1}(1) f_{j,q_2}(1) \eta_{i,t-q_1} \eta_{j,t-q_2} \\
&\triangleq U_{11}(T) + U_{12}(T).
\end{aligned} \tag{A.71}$$

Note that when $d_{ij} = 1$,

$$\begin{aligned}
\mathrm{E}[U_{11}(T)] &= \frac{\sigma_1^2}{T^{d_{ij}}} \sum_{t=1}^{T} \sum_{q_1=1}^{\infty} f_{i,q_1}(1) f_{j,q_1}(1) \mathrm{E}\left[\eta_{i,t-q_1} \eta_{j,t-q_1}\right] \\
&= \sigma_1^2 \sigma_{ij} \sum_{q_1=1}^{\infty} f_{i,q_1}(1) f_{j,q_1}(1)
\end{aligned} \tag{A.72}$$

and

$$\mathrm{E}[U_{12}(T)] = \frac{\sigma_1^2}{n} \sum_{t=1}^{T} \sum_{q_1=1}^{\infty} \sum_{\substack{q_2=1 \\ q_2 \neq q_1}}^{\infty} f_{i,q_1}(1) f_{j,q_2}(1) \mathrm{E}[\eta_{i,t-q_1} \eta_{j,t-q_2}] = 0. \tag{A.73}$$

Meanwhile, we have $Var[U_{11}(T)] = O(T^{-1})$ and $Var[U_{12}(T)] = O(T^{-1})$. Therefore, when $d_{ij} = 1$, as $T \to \infty$,

$$U_{11}(T) \longrightarrow_P \sigma_1^2 \sigma_{ij} \sum_{q_1=1}^{\infty} f_{i,q_1}(1) f_{j,q_1}(1), \tag{A.74}$$

$$U_{12}(T) \longrightarrow_P 0. \tag{A.75}$$

When $d_{ij} > 1$,

$$\begin{aligned} \mathrm{E}[U_{11}(T)] &= \frac{\sigma_1^2}{T^{d_{ij}}} \sum_{t=1}^{T} \sum_{q_1=1}^{\infty} f_{i,q_1}(1) f_{j,q_1}(1) \mathrm{E}\left[\eta_{i,t-q_1}\eta_{j,t-q_1}\right] \\ &= T^{d_{ij}-1} \sigma_1^2 \sigma_{ij} \sum_{q_1=1}^{\infty} f_{i,q_1}(1) f_{j,q_1}(1) \longrightarrow 0 \end{aligned} \tag{A.76}$$

and

$$\mathrm{E}[U_{12}(T)] = \frac{\sigma_1^2}{T^{d_{ij}}} \sum_{t=1}^{T} \sum_{q_1=1}^{\infty} \sum_{\substack{q_2=1 \\ q_2 \neq q_1}}^{\infty} f_{i,q_1}(1) f_{j,q_2}(1) \mathrm{E}[\eta_{i,t-q_1}\eta_{j,t-q_2}] = 0. \tag{A.77}$$

Meanwhile, $Var[U_{11}(T)] = O(T^{-1})$ and $Var[U_{12}(T)] = O(T^{-2d_{ij}+1})$. Therefore, when $d_{ij} > 1$, as $T \to \infty$,

$$U_{11}(T) \longrightarrow_P 0, \tag{A.78}$$

$$U_{12}(T) \longrightarrow_P 0. \tag{A.79}$$

Hence, when $d_{ij} = 1$,

$$\sum_{t=1}^{T} \mathrm{E}[M_{3,Tt}^{i} M_{3,nt}^{j} | F_{t-1}] \longrightarrow_P \sigma_1^2 \sigma_{ij} \sum_{q_1=1}^{\infty} f_{i,q_1}(1) f_{j,q_1}(1), \tag{A.80}$$

when $d_{ij} > 1$,

$$\sum_{t=1}^{T} \mathrm{E}[M_{3,Tt}^{i} M_{3,nt}^{j} | F_{t-1}] \longrightarrow_P 0. \tag{A.81}$$

(5) When $r_1 = r_2 = 4$,

$$
\begin{aligned}
&\sum_{t=1}^{T} \mathrm{E}[M_{4,Tt}^{i} M_{4,nt}^{j} | F_{t-1}] \\
&\quad = \frac{1}{T^{d_{ij}}} \sum_{t=1}^{T} \mathrm{E}\left[\sum_{q_1=1}^{\infty} m_{i,q_1}(1) \epsilon_{t-q_1} \eta_{i,t} \sum_{q_2=1}^{\infty} m_{j,q_2}(1) \epsilon_{t-q_2} \eta_{j,t} | F_{t-1} \right] \\
&\quad = \frac{1}{T^{d_{ij}}} \sum_{t=1}^{T} \sum_{q_1=1}^{\infty} \sum_{q_2=1}^{\infty} m_{i,q_1}(1) m_{j,q_2}(1) \epsilon_{t-q_1} \epsilon_{t-q_2} \mathrm{E}\left[\eta_{i,t} \eta_{j,t}\right] \\
&\quad = \frac{\sigma_{ij}}{T^{d_{ij}}} \sum_{t=1}^{T} \sum_{q_1=1}^{\infty} \sum_{q_2=1}^{\infty} m_{i,q_1}(1) m_{j,q_2}(1) \epsilon_{t-q_1} \epsilon_{t-q_2} \\
&\quad = \frac{\sigma_{ij}}{T^{d_{ij}}} \sum_{t=1}^{T} \sum_{q_1=1}^{\infty} m_{i,q_1}(1) m_{j,q_1}(1) \epsilon_{t-q_1}^{2} \\
&\qquad + \frac{\sigma_{ij}}{T^{d_{ij}}} \sum_{t=1}^{T} \sum_{q_1=1}^{\infty} \sum_{\substack{q_2=1 \\ q_2 \neq q_1}}^{\infty} m_{i,q_1}(1) m_{j,q_2}(1) \epsilon_{t-q_1} \epsilon_{t-q_2} \\
&\quad \triangleq U_{21}(T) + U_{22}(T).
\end{aligned} \tag{A.82}
$$

It is easy to show that when $d_{ij} = 1$,

$$
\mathrm{E}[U_{21}(T)] = \sigma_{ij} \sigma_1^2 \sum_{q_1=1}^{\infty} m_{i,q_1}(1) m_{j,q_1}(1), \tag{A.83}
$$

$$
\mathrm{E}[U_{22}(T)] = 0. \tag{A.84}
$$

Using the same method as in the previous case, we can show that

$$
\mathrm{Var}[U_{21}(T)] \longrightarrow 0, \tag{A.85}
$$

$$
\mathrm{Var}[U_{22}(T)] \longrightarrow 0. \tag{A.86}
$$

Therefore, when $d_{ij} = 1$,

$$U_{21}(T) \longrightarrow_P \sigma_{ij}\sigma_1^2 \sum_{q_1=1}^{\infty} m_{i,q_1}(1)m_{j,q_1}(1), \tag{A.87}$$

$$U_{22}(T) \longrightarrow_P 0. \tag{A.88}$$

When $d_{ij} > 1$, we can show that

$$\mathrm{E}[U_{21}(T)] = 0, \mathrm{E}[U_{22}(T)] = 0, \tag{A.89}$$

$$\mathrm{Var}[U_{21}(T)] \longrightarrow 0, Var[U_{22}(T)] \longrightarrow 0. \tag{A.90}$$

Therefore, when $d_{ij} > 1$,

$$U_{21}(T) \longrightarrow_P 0, \tag{A.91}$$

$$U_{22}(T) \longrightarrow_P 0. \tag{A.92}$$

Hence, when $d_{ij} = 1$,

$$\sum_{t=1}^{T} \mathrm{E}[M_{4,Tt}^i M_{4,Tt}^j | F_{t-1}] \longrightarrow_P \sigma_{ij}\sigma_1^2 \sum_{q_1=1}^{\infty} m_{i,q_1}(1)m_{j,q_1}(1). \tag{A.93}$$

When $d_{ij} > 1$,

$$\sum_{t=1}^{T} \mathrm{E}[M_{4,Tt}^i M_{4,Tt}^j | F_{t-1}] \longrightarrow_P 0. \tag{A.94}$$

(6) When $r_1 = 1, r_2 = 3$ or $r_1 = 3, r_2 = 1$,

$$\sum_{t=1}^{T} \mathrm{E}[M_{1,Tt}^i M_{3,Tt}^j | F_{t-1}] = \frac{1}{T^{d_{ij}}} \sum_{t=1}^{T} \mathrm{E}\left[g_i(\tau_t)\Phi(1)\epsilon_t \sum_{q=1}^{\infty} f_{j,q}(1)\epsilon_t \eta_{j,t-q} | F_{t-1} \right]$$

$$= \frac{\sigma_1^2 \Phi(1)}{T^{d_{ij}}} \sum_{t=1}^{T} g_i(\tau_t) \sum_{q=1}^{\infty} f_{j,q}(1)\eta_{j,t-q} \triangleq U_3(T). \tag{A.95}$$

Note that $\mathrm{E}[U_3(T)] = 0$ and

$$\mathrm{E}[U_3(T)^2] = \frac{\sigma_1^4\Phi(1)^2}{T^{2d_{ij}}}\mathrm{E}\left[\left(\sum_{t=1}^{T} g_i(\tau_t)\sum_{q=1}^{\infty} f_{j,q}(1)\eta_{j,t-q}\right)^2\right]$$

$$= \frac{\sigma_1^4\Phi(1)^2\sigma_{jj}}{T^{2d_{ij}}}\sum_{t_1=1}^{T}\sum_{q_1=1}^{\infty} g_i(\tau_{t_1})^2 f_{j,q_1}(1)^2$$

$$+ \frac{2\sigma_1^4\Phi(1)^2\sigma_{jj}}{T^{2d_{ij}}}\sum_{t_1=1}^{T-1}\sum_{t_2=t_1+1}^{T}\sum_{q_1=1}^{\infty} g_i(\tau_{t_1})g_i(\tau_{t_2})f_{j,q_1}(1)f_{j,t_2-t_1+q_1}(1), \tag{A.96}$$

where the first term is $O(T^{-d_j})$ given that $\sum_{q_1=1}^{\infty} f_{j,q_1}(1)^2 < \infty$. For the second term

$$\frac{1}{T^{2d_{ij}-1}}\sum_{t_1=1}^{T-1}\sum_{t_2=t_1+1}^{T}\sum_{q_1=1}^{\infty} g_i(\tau_{t_1})g_i(\tau_{t_2})f_{j,q_1}(1)f_{j,t_2-t_1+q_1}(1)$$

$$= \frac{1}{T^{d_j-1}}\int_0^1\int_{\tau_1}^1 \frac{g_i(\tau_1)g_i(\tau_2)}{T^{d_i-1}}\gamma_2(T(\tau_2-\tau_1),j)d\tau_1 d\tau_2 \longrightarrow 0, \tag{A.97}$$

given that $\gamma_2(d,j) = \sum_{d=1}^{\infty}\sum_{q_1=1}^{\infty} f_{j,q_1}(1)f_{j,d+q_1}(1)$ and

$$\lim_{T\to\infty}\frac{1}{T^{d_j-1}}\int_0^1\int_{\tau_1}^1 \frac{g_i(\tau_1)g_i(\tau_2)}{T^{d_i-1}}\gamma_2(T(\tau_2-\tau_1),j)d\tau_1 d\tau_2 = 0. \tag{A.98}$$

Therefore, as $T \to \infty$,

$$\sum_{t=1}^{T}\mathrm{E}[M_{1,Tt}^i M_{3,Tt}^j | F_{t-1}] \longrightarrow_P 0. \tag{A.99}$$

A similar result holds when $r_1 = 3$ and $r_2 = 1$.
(7) When $r_1 = 1, r_2 = 4$ or $r_1 = 4, r_2 = 1$,

$$\sum_{t=1}^{T}\mathrm{E}[M_{1,Tt}^i M_{4,Tt}^j | F_{t-1}]$$

$$= \frac{1}{T^{d_{ij}}}\sum_{t=1}^{T}\mathrm{E}\left[g_i(\tau_t)\Phi(1)\epsilon_t\sum_{q=1}^{\infty} m_{j,q}(1)\epsilon_{t-q}\eta_{j,t}|F_{t-1}\right]$$

$$= \frac{\theta_j\Phi(1)}{T^{d_{ij}}}\sum_{t=1}^{T} g_i(\tau_t)\sum_{q=1}^{\infty} m_{j,q}(1)\epsilon_{t-q} \triangleq U_4(T). \tag{A.100}$$

We can show that $\mathrm{E}[U_4(T)] = 0$, and

$$\begin{aligned}
\mathrm{E}[U_4(T)^2] &= \frac{\theta_i^2\Phi(1)^2}{T^{2d_{ij}}}\mathrm{E}\left[\left(\sum_{t=1}^{T} g_i(\tau_t)\sum_{q=1}^{\infty} m_{j,q}(1)\epsilon_{t-q}\right)^2\right] \\
&= \frac{\theta_i^2\Phi(1)^2\sigma_1^2}{T^{2d_{ij}}}\sum_{t_1=1}^{n}\sum_{q_1=1}^{\infty} g_i(\tau_{t_1})^2 m_{j,q_1}(1)^2 \\
&\quad + \frac{\theta_i^2\Phi(1)^2\sigma_1^2}{T^{2d_{ij}}}\sum_{t_1=1}^{n-1}\sum_{t_2=t_1+1}^{n}\sum_{q_1=1}^{\infty} \\
&= g_i(\tau_{t_1})g_i(\tau_{t_2})m_{j,q_1}(1)m_{j,t_2-t_1+q_1}(1), \qquad \text{(A.101)}
\end{aligned}$$

where the first term is $O(T^{-d_j})$ given that $\sum_{q_1=1}^{\infty} m_{j,q_1}(1)^2 < \infty$. For the second term

$$\begin{aligned}
&\frac{1}{T^{2d_{ij}}}\sum_{t_1=1}^{T-1}\sum_{t_2=t_1+1}^{T}\sum_{q_1=1}^{\infty} g_i(\tau_{t_1})g_i(\tau_{t_2})m_{j,q_1}(1)m_{j,t_2-t_1+q_1}(1) \\
&= \frac{1}{T^{d_j-1}}\int_0^1\int_{\tau_1}^1 \frac{g_i(\tau_1)g_i(\tau_2)}{T^{d_i-1}}\gamma_3(T(\tau_2-\tau_1),j)d\tau_1 d\tau_2 \longrightarrow 0, \\
&\qquad \text{(A.102)}
\end{aligned}$$

where $\gamma_3(d,j) = \sum_{q_1=1}^{\infty} m_{j,q_1}(1)m_{j,d+q_1}(1)$ and

$$\lim_{T\to\infty}\frac{1}{T^{d_j-1}}\int_0^1\int_{\tau_1}^1 \frac{g_i(\tau_1)g_i(\tau_2)}{T^{d_i-1}}\gamma_3(T(\tau_2-\tau_1),j)d\tau_1 d\tau_2 = 0. \quad \text{(A.103)}$$

Therefore, as $T \to \infty$,

$$\sum_{t=1}^{T} \mathrm{E}[M_{1,Tt}^i M_{4,Tt}^j | F_{t-1}] \longrightarrow_P 0. \qquad \text{(A.104)}$$

A similar result holds when $r_1 = 4$ and $r_2 = 1$.

(8) When $r_1 = 2, r_2 = 3$ or $r_1 = 3, r_2 = 2$,

$$\begin{aligned}
&\sum_{t=1}^{T} \mathrm{E}[M_{2,Tt}^{i} M_{3,Tt}^{j} | F_{t-1}] \\
&\quad = \frac{1}{T^{d_{ij}}} \sum_{t=1}^{T} \mathrm{E}\left[f_{i,0}(1)\left(\epsilon_t \eta_{it} - \theta_i\right) \sum_{q=1}^{\infty} f_{j,q}(1) \epsilon_t \eta_{j,t-q} \middle| F_{t-1} \right] \\
&\quad = \frac{1}{T^{d_{ij}}} \sum_{t=1}^{T} f_{i,0}(1) \sum_{q=1}^{\infty} f_{j,q}(1) \eta_{j,t-q} \mathrm{E}\left[\left(\epsilon_t^2 \eta_{it} \right) \right] \\
&\quad = \frac{\delta_{2i} f_{i,0}(1)}{T^{d_{ij}}} \sum_{t=1}^{T} \sum_{q=1}^{\infty} f_{j,q}(1) \eta_{j,t-q} \triangleq U_5(T).
\end{aligned} \tag{A.105}$$

It is obvious that $\mathrm{E}[U_5(T)] = 0$, and

$$\begin{aligned}
\mathrm{E}[U_5(T)^2] &= \mathrm{E}\left[\left(\frac{\delta_{2i} f_{i,0}(1)}{T^{d_{ij}}} \sum_{t=1}^{T} \sum_{q=1}^{\infty} f_{j,q}(1) \eta_{j,t-q} \right)^2 \right] \\
&= \frac{\delta_{2i}^2 f_{i,0}(1)^2}{T^{2d_{ij}}} \sum_{t_1=1}^{T} \sum_{t_2=1}^{T} \sum_{q_1=1}^{\infty} \sum_{q_2=1}^{\infty} f_{j,q_1}(1) f_{j,q_2}(1) \mathrm{E}\left[\eta_{j,t_1-q_1} \eta_{j,t_2-q_2} \right] \\
&= \frac{\delta_{2i}^2 f_{i,0}(1)^2 \sigma_{jj}}{T^{2d_{ij}}} \sum_{t_1=1}^{T} \sum_{q_1=1}^{\infty} f_{j,q_1}(1)^2 \\
&\quad + \frac{\delta_{2i}^2 f_{i,0}(1)^2 \sigma_{jj}}{T^{2d_{ij}}} \sum_{t_1=1}^{T-1} \sum_{t_2=t_1+1}^{T} \sum_{q_1=1}^{\infty} f_{j,q_1}(1) f_{j,t_2-t_1+q_1}(1) \\
&= O(T^{-2d_{ij}+1}),
\end{aligned} \tag{A.106}$$

given that $\sum_{q_1=1}^{\infty} f_{j,q_1}(1)^2 < \infty$ and $\sum_{p=1}^{\infty} \sum_{q_1=1}^{\infty} f_{j,q_1}(1) f_{j,p+q_1}(1) < \infty$. Therefore,

$$\sum_{t=1}^{T} \mathrm{E}[M_{2,Tt}^{i} M_{3,Tt}^{j} | F_{t-1}] \longrightarrow_P 0. \tag{A.107}$$

A similar result holds when $r_1 = 3$ and $r_2 = 2$.

(9) When $r_1 = 2, r_2 = 4$ or $r_1 = 4, r_2 = 2$,

$$\sum_{t=1}^{T} \mathrm{E}[M_{2,Tt}^{i} M_{4,Tt}^{j} | F_{t-1}]$$

$$= \frac{1}{T^{d_{ij}}} \sum_{t=1}^{T} \mathrm{E}\left[f_{i,0}(1) \left(\epsilon_t \eta_{it} - \theta_i\right) \sum_{q=1}^{\infty} m_{j,q}(1) \epsilon_{t-q} \eta_{j,t} \middle| F_{t-1} \right]$$

$$= \frac{f_{i,0}(1)}{T^{d_{ij}}} \sum_{t=1}^{T} \sum_{q=1}^{\infty} m_{j,q}(1) \epsilon_{t-q} \mathrm{E}[\epsilon_t \eta_{it} \eta_{j,t}]$$

$$= \frac{f_{i,0}(1)\delta_{1ij}}{T^{d_{ij}}} \sum_{t=1}^{T} \sum_{q=1}^{\infty} m_{j,q}(1) \epsilon_{t-q} = U_6(T), \tag{A.108}$$

where $\delta_{1ij} = \mathrm{E}[\epsilon_t \eta_{it} \eta_{j,t}]$. Similar to the previous case, we can show that $\mathrm{E}[U_6(T)] = 0$, and

$$\mathrm{E}\left[U_6(T)^2\right] = \mathrm{E}\left[\left(\frac{f_{i,0}(1)\delta_{1ij}}{T^{d_{ij}}} \sum_{t=1}^{T} \sum_{q=1}^{\infty} m_{j,q}(1) \epsilon_{t-q} \right)^2 \right]$$

$$= \frac{f_{i,0}(1)^2 \delta_{1ij}^2}{T^{2d_{ij}}} \sum_{t=1}^{T} \sum_{q=1}^{\infty} m_{j,q}(1)^2 \mathrm{E}\left[\epsilon_{t-q}^2\right]$$

$$+ \frac{f_{i,0}(1)^2 \delta_{1ij}^2}{T^{2d_{ij}}} \sum_{t_1=1}^{T-1} \sum_{t_2=t_1+1}^{T} \sum_{q=1}^{\infty} m_{j,q}(1) m_{j,t_2-t_1+q} \mathrm{E}\left[\epsilon_{t_1-q}^2\right]$$

$$= \frac{f_{i,0}(1)^2 \delta_{1ij}^2 \sigma_1^2}{T^{2d_{ij}}} \sum_{t=1}^{T} \sum_{q=1}^{\infty} m_{j,q}(1)^2$$

$$+ \frac{f_{i,0}(1)^2 \delta_{1ij}^2 \sigma_1^2}{T^{2d_{ij}}} \sum_{t_1=1}^{T-1} \sum_{p=1}^{T-t_1} \sum_{q=1}^{\infty} m_{j,q}(1) m_{j,p+q}$$

$$= O(T^{-2d_{ij}+1}), \tag{A.109}$$

given that $\sum_{q=1}^{\infty} m_{j,q}(1)^2 < \infty$ and $\sum_{p=1}^{\infty} \sum_{q=1}^{\infty} m_{j,q}(1) m_{j,p+q} < \infty$.

Therefore, as $T \to \infty$,

$$\sum_{t=1}^{T} \mathrm{E}[M_{2,Tt}^{i} M_{4,Tt}^{j} | F_{t-1}] \longrightarrow_P 0. \tag{A.110}$$

A similar result holds when $r_1 = 4$ and $r_2 = 2$.
(10) When $r_1 = 3, r_2 = 4$ or $r_1 = 4, r_2 = 3$,

$$\begin{aligned}
&\sum_{t=1}^{T} \mathrm{E}[M_{3,Tt}^{i} M_{4,Tt}^{j} | F_{t-1}] \\
&\quad = \frac{1}{T^{d_{ij}}} \sum_{t=1}^{T} \mathrm{E}\left[\sum_{q=1}^{\infty} f_{i,q}(1) \epsilon_t \eta_{i,t-q} \sum_{l=1}^{\infty} m_{j,l}(1) \epsilon_{t-l} \eta_{j,t} \middle| F_{t-1} \right] \\
&\quad = \frac{1}{T^{d_{ij}}} \sum_{t=1}^{T} \sum_{q=1}^{\infty} \sum_{l=1}^{\infty} f_{i,q}(1) \eta_{i,t-q} m_{j,l}(1) \epsilon_{t-l} \mathrm{E}\left[\epsilon_t \eta_{j,t} \right] \\
&\quad = \frac{\theta_j}{T^{d_{ij}}} \sum_{t=1}^{T} \sum_{q=1}^{\infty} \sum_{l=1}^{\infty} f_{i,q}(1) m_{j,l}(1) \eta_{i,t-q} \epsilon_{t-l} \\
&\quad = \frac{\theta_j}{T^{d_{ij}}} \sum_{t=1}^{T} \sum_{q=1}^{\infty} f_{i,q}(1) m_{j,q}(1) \eta_{i,t-q} \epsilon_{t-q} \\
&\qquad + \frac{\theta_j}{T^{d_{ij}}} \sum_{t=1}^{T} \sum_{q=1}^{\infty} \sum_{l=q+1}^{\infty} f_{i,q}(1) m_{j,l}(1) \eta_{i,t-q} \epsilon_{t-l} \\
&\quad \triangleq U_{71}(T) + U_{72}(T).
\end{aligned} \tag{A.111}$$

Then, when $d_{ij} = 1$,

$$\mathrm{E}[U_{71}(T)] = \mathrm{E}\left[\frac{\theta_j}{T} \sum_{t=1}^{T} \sum_{q=1}^{\infty} f_{i,q}(1) m_{j,q}(1) \eta_{i,t-q} \epsilon_{t-q} \right] = \theta_j \theta_i \sum_{q=1}^{\infty} f_{i,q}(1) m_{j,q}(1) \tag{A.112}$$

and

$$\mathrm{E}[U_{72}(T)] = \frac{\theta_j}{T} \sum_{t=1}^{T} \sum_{q=1}^{\infty} \sum_{l=q+1}^{\infty} f_{i,q}(1) m_{j,l}(1) \mathrm{E}\left[\eta_{i,t-q} \epsilon_{t-l} \right] = 0. \tag{A.113}$$

We can show that

$$\mathrm{E}\Big[(U_{71}(T) - \mathrm{E}[U_{71}(T)])^2 \Big] = O(T^{-1}) \tag{A.114}$$

and

$$\mathrm{E}[U_{72}(T)^2] = O(T^{-1}). \tag{A.115}$$

Therefore, when $d_{ij} = 1$, as $T \to \infty$,

$$U_{71}(T) \longrightarrow_P \theta_j \theta_i \sum_{q=1}^{\infty} f_{i,q}(1) m_{j,q}(1), \tag{A.116}$$

$$U_{72}(T) \longrightarrow_P 0. \tag{A.117}$$

Hence,

$$\sum_{t=1}^{T} \mathrm{E}[M^i_{3,Tt} M^j_{4,Tt} | F_{t-1}] = U_{71}(T) + U_{72}(T) \longrightarrow_P \theta_j \theta_i \sum_{q=1}^{\infty} f_{i,q}(1) m_{j,q}(1). \tag{A.118}$$

When $d_{ij} > 1$,

$$\begin{aligned} \mathrm{E}[U_{71}(T)] &= \mathrm{E}\left[\frac{\theta_j}{T^{d_{ij}}} \sum_{t=1}^{T} \sum_{q=1}^{\infty} f_{i,q}(1) m_{j,q}(1) \eta_{i,t-q} \epsilon_{t-q} \right] \\ &= \frac{\theta_j \theta_i}{T^{d_{ij}-1}} \sum_{q=1}^{\infty} f_{i,q}(1) m_{j,q}(1) \longrightarrow 0 \end{aligned} \tag{A.119}$$

and

$$\mathrm{E}[U_{72}(T)] = \frac{\theta_j}{T^{d_{ij}}} \sum_{t=1}^{T} \sum_{q=1}^{\infty} \sum_{l=q+1}^{\infty} f_{i,q}(1) m_{j,l}(1) \mathrm{E}\left[\eta_{i,t-q} \epsilon_{t-l} \right] = 0. \tag{A.120}$$

Meanwhile,

$$\mathrm{E}\Big[U_{71}(T)^2 \Big] = O(T^{-2d_{ij}+1}) \tag{A.121}$$

and

$$\mathrm{E}[U_{72}(T)^2] = O(T^{-2d_{ij}+1}). \tag{A.122}$$

Therefore, when $d_{ij} = 1$, as $T \to \infty$,

$$U_{71}(T) \longrightarrow_P 0, \tag{A.123}$$

$$U_{72}(T) \longrightarrow_P 0. \tag{A.124}$$

Hence, when $d_{ij} = 1$,

$$\sum_{t=1}^{T} \mathrm{E}[M_{3,Tt}^{i} M_{4,Tt}^{j} | F_{t-1}] = U_{71}(T) + U_{72}(T) \longrightarrow_P \theta_j \theta_i \sum_{q=1}^{\infty} f_{i,q}(1) m_{j,q}(1). \tag{A.125}$$

When $d_{ij} > 1$,

$$\sum_{t=1}^{T} \mathrm{E}[M_{3,Tt}^{i} M_{4,Tt}^{j} | F_{t-1}] = U_{71}(T) + U_{72}(T) \longrightarrow_P 0. \tag{A.126}$$

A similar result holds when $r_1 = 4$ and $r_2 = 3$.
To conclude,

$$\sum_{t=1}^{T} \mathrm{E}\left[(a' M_{Tt})^2 | F_{t-1}\right] \longrightarrow_P a' \Omega a, \tag{A.127}$$

where Ω is a $K \times K$ variance–covariance matrix defined as follows:
For $1 \leq i \leq K_1$ and $1 \leq j \leq K_1$, i.e., $d_i = d_j = 1$,

$$\begin{aligned}
\Omega_{ij} &= \sigma_1^2 \Phi(1)^2 \boldsymbol{Q}_{ij} + f_{i,0}(1) f_{j,0}(1)(\delta_{2ij} - \theta_i \theta_j) \\
&\quad + \Phi(1) f_{j,0}(1) \delta_{2j} \bar{g}_i + \Phi(1) f_{i,0}(1) \delta_{2i} \bar{g}_j \\
&\quad + \sigma_1^2 \sigma_{ij} \sum_{q_1=1}^{\infty} f_{i,q_1}(1) f_{j,q_1}(1) + \sigma_1^2 \sigma_{ij} \sum_{q_1=1}^{\infty} m_{i,q_1}(1) m_{j,q_1}(1) \\
&\quad + \theta_j \theta_i \sum_{q=1}^{\infty} f_{i,q}(1) m_{j,q}(1) + \theta_i \theta_j \sum_{q=1}^{\infty} f_{j,q}(1) m_{i,q}(1).
\end{aligned} \tag{A.128}$$

For $K_1 < i \leq n$ and $1 \leq j \leq K_1$, i.e., $d_i > 1$, $d_j = 1$,

$$\Omega_{ij} = \sigma_1^2 \Phi(1)^2 \boldsymbol{Q}_{ij} + \Phi(1) f_{j,0}(1) \delta_{2j} \bar{g}_i. \tag{A.129}$$

Finally, when $K_1 < i \leq K$ and $K_1 < j \leq K$, i.e., $d_i > 1$, $d_j > 1$,

$$\Omega_{ij} = \sigma_1^2 \Phi(1)^2 \boldsymbol{Q}_{ij}. \tag{A.130}$$

We then examine the second condition of the CLT for martingale difference sequence that as $T \to \infty$,

$$\sum_{t=1}^{T} \mathrm{E}\left[(a' M_{Tt})^4 \middle| F_{t-1} \right] \longrightarrow_P 0. \tag{A.131}$$

It is then equivalent to show that for any i,

$$\sum_{t=1}^{T} \mathrm{E}\left[a_i^4 (M_{Tt}^i)^4 \middle| F_{t-1} \right] \longrightarrow_P 0. \tag{A.132}$$

Then, it is equivalent to prove

$$\sum_{t=1}^{T} \mathrm{E}\left[(M_{p,Tt}^i)^4 \middle| F_{t-1} \right] \longrightarrow_P 0, \tag{A.133}$$

for $p = 1, 2, 3, 4$, respectively.
When $p = 1$,

$$\begin{aligned} \sum_{t=1}^{T} \mathrm{E}\left[(M_{1,Tt}^i)^4 | F_{t-1} \right] &= \frac{1}{T^{2d_i}} \sum_{t=1}^{T} g_i(\tau_t)^4 \Phi(1)^4 \mathrm{E}[\epsilon_t^4] \\ &= \frac{C\Phi(1)^4}{T^{2d_i}} \sum_{t=1}^{T} g_i(\tau_t)^4, \end{aligned} \tag{A.134}$$

which is purely deterministic. Note that

$$\frac{1}{T} \sum_{t=1}^{T} \left(T^{-\frac{d_i - 1}{2}} g_i(\tau_t) \right)^4 \longrightarrow \int_0^1 g_i^N(\tau)^4 d\tau < \infty,$$

where $g_i^N(\tau) = T^{-\frac{d_i - 1}{2}} g_i(\tau_t)$ is the rescaled trend function. Therefore, given that $\mathrm{E}[\epsilon_t^4] < \infty$, we have as $T \to \infty$

$$\sum_{t=1}^{T} \mathrm{E}\left[(M_{1,Tt}^i)^4 | F_{t-1} \right] = \frac{1}{T} \left(C\Phi(1)^4 \int_0^1 g_i^N(\tau) d\tau \right) \longrightarrow 0. \tag{A.135}$$

When $p = 2$,

$$\sum_{t=1}^{T} \mathrm{E}\left[(M_{2,Tt}^{i})^4|F_{t-1}\right] = \frac{1}{T^{2d_i}} \sum_{t=1}^{T} f_{i,0}(1)^4 \mathrm{E}\left[(\epsilon_t \eta_{it} - \sigma_{12})^4\right]$$
$$= \frac{(\delta_{44i} - \sigma_{12}^4)}{T^{2d_i}} \sum_{t=1}^{T} f_{i,0}(1)^4$$
$$= \frac{(\delta_{44i} - \sigma_{12}^4) f_{i,0}(1)^4}{T^{2d_i - 1}}, \tag{A.136}$$

where $\delta_{44i} = \mathrm{E}[\epsilon_t^4 \eta_{it}^4] < \infty$. Since it is also purely deterministic and $2d_i - 1 > 0$, therefore, as $T \to \infty$

$$\sum_{t=1}^{T} \mathrm{E}\left[(M_{2,Tt}^{i})^4|F_{t-1}\right] \longrightarrow 0. \tag{A.137}$$

When $p = 3$,

$$\sum_{t=1}^{T} \mathrm{E}\left[(M_{3,Tt}^{i})^4|F_{t-1}\right] = \frac{1}{T^{2d_i}} \sum_{t=1}^{T} \left(\sum_{q=1}^{\infty} f_{i,q}(1)\eta_{i,t-q}\right)^4 \mathrm{E}[\epsilon_t^4]$$
$$= \frac{C}{T^{2d_i}} \sum_{t=1}^{T} \left(\sum_{q=1}^{\infty} f_{i,q}(1)\eta_{i,t-q}\right)^4 \triangleq C \cdot U_8(T). \tag{A.138}$$

Note that

$$\mathrm{E}[U_8(T)]$$
$$= \frac{C}{T^{2d_i - 1}} \sum_{k=1}^{\infty} f_{i,q}(1)^4 + \frac{C}{T^{2d_i - 1}} \sum_{q_1=1}^{\infty} \sum_{q_2=q_1+1}^{\infty} f_{q_1}(1)^2 f_{q_2}(1)^2$$
$$= O(T^{-2d_i+1}) = o(1), \tag{A.139}$$

since $2d_i - 1 > 0$ and given that $\mathrm{E}\left[\eta_{i,t}^4\right] < \infty$, $\mathrm{E}\left[\eta_{i,t-q_1}^2 \eta_{i,t-q_2}^2\right] < \infty$ and $\sum_{q=1}^{\infty} f_{i,q}(1)^4 < \infty$, $\sum_{q_1=1}^{\infty} \sum_{q_2=q_1+1}^{\infty} f_{i,q_1}(1)^2 f_{i,q_2}(1)^2 < \infty$.

Since $U_8(T) \geq 0$, then as $T \to \infty$, $\mathrm{E}[U_8(T)] \longrightarrow 0$ implies

$$U_8(T) \longrightarrow_P 0. \tag{A.140}$$

Using the same method as above, we can show that when $p = 4$,

$$\begin{aligned}\sum_{t=1}^{T} \mathrm{E}\left[(M^i_{4,Tt})^4 | F_{t-1}\right] &= \frac{1}{T^{2d_i}} \sum_{t=1}^{T} \left(\sum_{q=1}^{\infty} m_{i,q}(1)\epsilon_{t-q}\right)^4 \mathrm{E}[\eta^4_{it}] \\ &= \frac{C}{T^{2d_i}} \sum_{t=1}^{T} \left(\sum_{q=1}^{\infty} m_{i,q}(1)\epsilon_{t-q}\right)^4 \triangleq C \cdot U_9(T).\end{aligned} \tag{A.141}$$

Note that

$$\begin{aligned}&\mathrm{E}[U_9(T)] \\ &= \frac{C}{T^{2d_i-1}} \sum_{q_1=1}^{\infty} m_{i,q_1}(1)^4 + \frac{C}{T^{2d_i-1}} \sum_{q_1=1}^{\infty} \sum_{q_2=q_1+1}^{\infty} m_{i,q_1}(1)^2 m_{i,q_2}(1)^2 \\ &= O(T^{-2d_i+1}) = o(1),\end{aligned} \tag{A.142}$$

given that $\mathrm{E}[\epsilon^4_{t-q_1}] < \infty$ and $\mathrm{E}[\epsilon^2_{t-q_1}\epsilon^2_{t-q_2}] < \infty$. Meanwhile, $\sum_{q=1}^{\infty} m_{i,q}(1)^4 < \infty$ and $\sum_{q_1=1}^{\infty} \sum_{q_2=q_1+1}^{\infty} m_{i,q_1}(1)^2 m_{i,q_2}(1)^2 < \infty$. Hence, as $U_9(T) \geq 0$, $\mathrm{E}[U_9(T)] \longrightarrow 0$ implies

$$U_9(T) \longrightarrow_P 0, \tag{A.143}$$

as $T \to \infty$. Therefore, equation (A.132) holds. We then complete the proof for Lemma 4.8.5.

References

Amihud, Y. and Hurvich, C.M. (2004). Predictive regressions: A reduced-bias estimation method, *Journal of Financial and Quantitative Analysis*, **39**(4): 813–841.

Amihud, Y., Hurvich, C.M. and Wang, Y. (2009). Multiple-predictor regressions: Hypothesis testing, *The Review of Financial Studies*, **22**(1): 413–434.

Amihud, Y., Hurvich, C.M. and Wang, Y. (2010). Predictive regression with order-p autoregressive predictors, *Journal of Empirical Finance*, **17**(3): 513–525.

Andersen, T.G. and Varneskov, R.T. (2021). Consistent inference for predictive regressions in persistent economic systems, *Journal of Econometrics*, **224**(1): 215–244.

Anderson, H.M. and Vahid, F. (1998). Testing multiple equation systems for common nonlinear components, *Journal of Econometrics*, **84**(1): 1–36.

Andreas, E.L. and Treviño, G. (1997). Using wavelets to detect trends, *Journal of Atmospheric and Oceanic Technology*, **14**(3): 554–564.

Andrews, D.W.K. (1991). Heteroskedasticity and autocorrelation consistent co-variance matrix estimation, *Econometrica*, **59**(3): 817–854.

Angrist, J.D., Jordà, Ò. and Kuersteiner, G.M. (2018). Semiparametric estimates of monetary policy effects: String theory revisited, *Journal of Business and Economic Statistics*, **36**(3): 371–387.

Bae, Y. and De Jong, R.M. (2007). Money demand function estimation by nonlinear cointegration, *Journal of Applied Econometrics*, **22**(4): 767–793.

Bengtsson, L., Hagemann, S. and Hodges, K.I. (2004). Can climate trends be calculated from reanalysis data? *Journal of Geophysical Research: Atmospheres*, **109:** D11.

Bernanke, B. (1985). Adjustment costs, durables, and aggregate consumption, *Journal of Monetary Economics*, **15**(1): 41–68.

Beveridge, S. and Nelson, C.R. (1981). A new approach to decomposition of economic time series into permanent and transitory components with particular attention to measurement of the 'business cycle', *Journal of Monetary Economics*, **7**(2): 151–174.

Bierens, H.J. (1997). Testing the unit root with drift hypothesis against nonlinear trend stationarity, with an application to the US price level and interest rate, *Journal of Econometrics*, **81**(1): 29–64.

Bierens, H.J. (2000). Nonparametric nonlinear cotrending analysis, with an application to interest and inflation in the United States, *Journal of Business and Economic Statistics*, **18**(3): 323–337.

Bierens, H.J. and Martins, L.F. (2010). Time-varying cointegration, *Econometric Theory*, **26**(5): 1453–1490.

Billingsley, P. (1999). *Convergence of Probability Measures*, (2nd ed.). John Wiley & Sons, New York.

Boneva, L., Linton, O. and Vogt, M. (2015). A semiparametric model for heterogeneous panel data with fixed effects, *Journal of Econometrics*, **188**(2): 327–345.

Box, G.E., Pierce, D.A. and Newbold, P. (1987). Estimating trend and growth rates in seasonal time series, *Journal of the American Statistical Association*, **82**(397): 276–282.

Breitung, J. (2002). Nonparametric tests for unit roots and cointegration, *Journal of Econometrics*, **108**(2): 343–363.

Breusch, T. and Vahid, F. (2011). Global temperature trends, *Report Provided for the Garnaut Climate Change Review*. Department of Econometrics and Business Statistics Working Paper 4-11.

Brown, J.P., Song, H. and McGillivray, A. (1997). Forecasting UK house prices: A time varying coefficient approach, *Economic Modelling*, **14**(4): 529–548.

Bruns, S.B., Csereklyei, Z. and Stern, D.I. (2020). A multicointegration model of global climate change, *Journal of Econometrics*, **214**(1): 175–197.

Bunzel, H. and Vogelsang, T.J. (2005). Powerful trend function tests that are robust to strong serial correlation, with an application to the Prebisch–Singer hypothesis, *Journal of Business and Economic Statistics*, **23**(4): 381–394.

Burke, M., Hsiang, S.M. and Miguel, E. (2015). Global non-linear effect of temperature on economic production, *Nature*, **527**(7577): 235–239.

Busetti, F. and Harvey, A. (2008). Testing for trend, *Econometric Theory*, **24**(1): 72–87.

Cai, B., Gao, J. and Tjøstheim, D. (2017). A new class of bivariate threshold cointegration models, *Journal of Business and Economic Statistics*, **35**(2): 288–305.

Cai, Z. (2007). Trending time-varying coefficient time series models with serially correlated errors, *Journal of Econometrics*, **136**(1): 163–188.

Cai, Z. and Wang, Y. (2014). Testing predictive regression models with nonstationary regressors, *Journal of Econometrics*, **178**(1): 4–14.

Cai, Z., Chen, L. and Fang, Y. (2018). A semiparametric quantile panel data model with an application to estimating the growth effect of FDI, *Journal of Econometrics*, **206**(2): 531–553.

Cai, Z., Fan, J. and Li, R. (2000). Efficient estimation and inferences for varying-coefficient models, *Journal of the American Statistical Association*, **95**(451): 888–902.

Cai, Z., Li, Q. and Park, J.Y. (2009). Functional-coefficient models for nonstationary time series data, *Journal of Econometrics*, **148**(2): 101–113.

Cai, Z., Ren, Y. and Yang, B. (2015). A semiparametric conditional capital asset pricing model, *Journal of Banking and Finance*, **61**: 117–126.

Campbell, J.Y. (1987). Does saving anticipate declining labor income? An alternative test of the permanent income hypothesis, *Econometrica*, **55**(6): 1249–1273.

Campbell, J.Y. and Shiller, R.J. (1988). The dividend-price ratio and expectations of future dividends and discount factors, *The Review of Financial Studies*, **1**(3): 195–228.

Campbell, J.Y. and Yogo, M. (2006). Efficient tests of stock return predictability, *Journal of Financial Economics*, **81**(1): 27–60.

Canjels, E. and Watson, M.W. (1997). Estimating deterministic trends in the presence of serially correlated errors, *The Review of Economics and Statistics*, **79**(2): 184–200.

Canova, F. (1998). Detrending and business cycle facts, *Journal of Monetary Economics*, **41**(3): 475–512.

Carleton, T.A. and Hsiang, S.M. (2016). Social and economic impacts of climate, *Science*, **353**(6304): aad9837.

Carson, M., Lyu, K., Richter, K., Becker, M., Domingues, C.M., Han, W. and Zanna, L. (2019). Climate model uncertainty and trend detection in regional sea level projections: A review, *Surveys in Geophysics*, **40**: 1631–1653.

Castle, J.L. and Hendry, D.F. (2022). Econometrics for modelling climate change: In *Oxford Research Encyclopedia of Economics and Finance*, Oxford University Press, Oxford.

Catania, L. and Luati, A. (2023). Semiparametric modeling of multiple quantiles, *Journal of Econometrics*, **237**(2): 105365.

Cavaliere, G., Harvey, D.I., Leybourne, S.J. and Taylor, A.R. (2011). Testing for unit roots in the presence of a possible break in trend and nonstationary volatility, *Econometric Theory*, **27**(5): 957–991.

Chan, N. and Wang, Q. (2015). Nonlinear regressions with nonstationary time series, *Journal of Econometrics*, **185**(1): 182–195.

Chang, Y. and Park, J.Y. (2011). Endogeneity in nonlinear regressions with integrated time series, *Econometric Reviews*, **30**(1): 51–87.

Chang, Y., Park, J.Y. and Phillips, P.C.B. (2001). Nonlinear econometric models with cointegrated and deterministically trending regressors, *The Econometrics Journal*, **4**(1): 1–36.

Chang, Y., Kaufmann, R.K., Kim, C.S., Miller, J.I., Park, J.Y. and Park, S. (2020). Evaluating trends in time series of distributions: A spatial fingerprint of human effects on climate, *Journal of Econometrics*, **214**(1): 274–294.

Chen, J., Gao, J. and Li, D. (2012). Semiparametric trending panel data models with cross-sectional dependence, *Journal of Econometrics*, **171**(1): 71–85.

Chen, J., Li, D., Linton, O. and Lu, Z. (2018). Semiparametric ultra-high dimensional model averaging of nonlinear dynamic time series, *Journal of the American Statistical Association*, **113**(522): 919–932.

Chen, L. (2017). *Trending Time Series Models with Endogeneity*, Ph.D. thesis, Department of Econometrics and Business Statistics, Monash University, Melbourne.

Chen, X., Liao, Z., and Sun, Y. (2014). Sieve inference on possibly misspecified semi-nonparametric time series models, *Journal of Econometrics*, **178**: 639–658.

Chen, Y. (2015). Semiparametric time series models with log-concave innovations: Maximum likelihood estimation and its consistency, *Scandinavian Journal of Statistics*, **42**(1): 1–31.

Chen, Y. and Tu, Y. (2019). Is stock price correlated with oil price? Spurious regressions with moderately explosive processes, *Oxford Bulletin of Economics and Statistics*, **81**(5): 1012–1044.

Choi, I. (2017). Efficient estimation of nonstationary factor models, *Journal of Statistical Planning and Inference*, **183**: 18–43.

Choi, I. and Saikkonen, P. (2010). Tests for nonlinear cointegration, *Econometric Theory*, **26**(3): 682–709.

Clements, M.P. and Hendry, D.F. (1999). *Forecasting Non-stationary Economic Time Series* MIT Press, Massachusetts.

Cochrane, J.H. (2012). A brief parable of over-differencing, Working Paper, University of Chicago, Booth School of Business.

Cogley, T., Primiceri, G.E. and Sargent, T.J. (2010). Inflation-gap persistence in the US, *American Economic Journal: Macroeconomics*, **2**(1): 43–69.

Cogley, T. and Sargent, T.J. (2005). Drifts and volatilities: Monetary policies and outcomes in the post WWII US, *Review of Economic Dynamics*, **8**(2): 262–302.

Cooley, T.F. and Prescott, E.C. (1976). Efficient estimation in the presence of stochastic parameter variation, *Econometrica*, **44**, 167–184.

Cubasch, U. and Meehl, G. (2001). Projections of future climate change. In *Climate Change 2001: The Scientific Basis*, Cambridge University Press, Cambridge, pp. 525–582.

Dangl, T. and Halling, M. (2012). Predictive regressions with time-varying coefficients, *Journal of Financial Economics*, **106**(1): 157–181.

Davidson, J. (2002). Establishing conditions for the functional central limit theorem in nonlinear and semiparametric time series processes, *Journal of Econometrics*, **106**(2): 243–269.

DeJong, D.N., Nankervis, J.C., Savin, N.E., and Whiteman, C.H. (1992). The power problems of unit root test in time series with autoregressive errors, *Journal of Econometrics*, **53**(1): 323–343.

DeJong, D.N. and Whiteman, C.H. (1991). Reconsidering 'trends and random walks in macroeconomic time series', *Journal of Monetary Economics*, **28**(2): 221–254.

De Jong, R.M. and Sakarya, N. (2016). The econometrics of the Hodrick-Prescott filter, *The Review of Economics and Statistics*, **98**(2): 310–317.

DeVore, R.A. and Lorentz, G.G. (1993). *Constructive Approximation*, Vol. 303, Springer, New York.

Dickey, D.A. and Fuller, W.A. (1979). Distribution of the estimators for autoregressive time series with a unit root, *Journal of the American Statistical Association*, **74**(366a): 427–431.

Diebold, F.X. and Rudebusch, G.D. (1991). On the power of Dickey-Fuller tests against fractional alternatives, *Economics Letters*, **35**(2): 155–160.

Dong, C. and Gao, J. (2018). Specification testing driven by orthogonal series for nonlinear cointegration with endogeneity, *Econometric Theory*, **34**(4): 754–789.

Dong, C. and Gao, J. (2019). Expansion of Lévy process functionals and its application in econometric estimation, *Econometric Reviews*, **39**: 125–150.

Dong, C., Gao, J., and Peng, B. (2015). Semiparametric single-index panel data models with cross-sectional dependence, *Journal of Econometrics*, **188**(1): 301–312.

Dong, C., Gao, J., Tjøstheim, D., and Yin, J. (2017). Specification testing for nonlinear multivariate cointegrating regressions, *Journal of Econometrics*, **200**(1): 104–117.

Dong, C. and Linton, O. (2018). Additive nonparametric models with time variable and both stationary and nonstationary regressors, *Journal of Econometrics*, **207**(1): 212–236.

Dong, C., Linton, O. and Peng, B. (2021). A weighted sieve estimator for nonparametric time series models with nonstationary variables, *Journal of Econometrics*, **222**(2): 909–932.

Düker, M.-C., Pipiras, V. and Sundararajan, R. (2022). Cotrending: Testing for common deterministic trends in varying means model, *Journal of Multivariate Analysis*, **187**: 104825.

Easterling, D.R., Evans, J., Groisman, P.Y., Karl, T.R., Kunkel, K.E. and Ambenje, P. (2000). Observed variability and trends in extreme climate events: A brief review, *Bulletin of the American Meteorological Society*, **81**(3): 417–426.

Elliott, G. (1998). On the robustness of cointegration methods when regressors almost have unit roots, *Econometrica*, **66**(1): 149–158.

Elliott, G. (2020). Testing for a trend with persistent errors, *Journal of Econometrics*, **219**(2): 314–328.

Elliott, G. and Timmermann, A. (2016). *Economic Forecasting*, Princeton University Press, New Jersey.

Elliott, G., Rothenberg, T.J. and Stock, J.H. (1996). Efficient tests for an autoregressive unit root, *Econometrica*, **64**(4): 813–836.

Engle, R.F. and Granger, C.W.J. (1987). Co-integration and error correction: Representation, estimation, and testing, *Econometrica*, **55**(2): 251–276.

Engle, R.F., Granger, C.W.J. and Hylleberg, S. (1993). Seasonal cointegration: The Japanese consumption function, *Journal of Econometrics*, **55**: 275–298.

Engle, R.F. and Hylleberg, S. (1996). Common seasonal features: Global unemployment, *Oxford Bulletin of Economics and Statistics*, **58**(4): 615–630.

Engle, R.F. and Kozicki, S. (1993). Testing for common features, *Journal of Business and Economic Statistics*, **11**(4): 369–380.

Estrada, F., Gay, C., and Sánchez, A. (2010). A reply to "Does temperature contain a stochastic trend? Evaluating conflicting statistical results" by R.K. Kaufmann *et al*, *Climatic Change*, **101**(3): 407–414.

Estrada, F., Perron, P., and Martínez-López, B. (2013). Statistically derived contributions of diverse human influences to twentieth-century temperature changes, *Nature Geoscience*, **6**(12): 1050–1055.

Fan, J. and Gijbels, I. (1996). *Local Polynomial Modelling and Its Applications: Monographs on Statistics and Applied Probability 66*, Chapman & Hall, London.

Fan, J. and Yao, Q. (2003). *Nonlinear Time Series: Nonparametric and Parametric Methods*, Springer, New York.

Fan, Y. and Li, Q. (1999). Central limit theorem for degenerate U-statistics of absolutely regular processes with applications to model specification testing, *Journal of Nonparametric Statistics*, **10**(3): 245–271.

Fan, Y. and Liu, R. (2016). A direct approach to inference in nonparametric and semiparametric quantile models, *Journal of Econometrics*, **191**(1): 196–216.

Favero, C.A., Gozluklu, A.E. and Tamoni, A. (2011). Demographic trends, the dividend-price ratio, and the predictability of long-run stock market returns, *Journal of Financial and Quantitative Analysis*, **46**(5): 1493–1520.

Flavin, M.A. (1981). The adjustment of consumption to changing expectations about future income, *The Journal of Political Economy*, **89**(5): 974–1009.

Flavin, M.A. (1984). Excess sensitivity of consumption to current income: Liquidity constraints or myopia? Working Paper, NBER Working Paper No. 1341.

Fu, Z., Hong, Y., Su, L., and Wang, X. (2023). Specification tests for time-varying coefficient models, *Journal of Econometrics*, **235**(2): 720–744.

Fuller, W.A. (1976). *Introduction to Statistical Time Series*, John Wiley & Sons, New York.

Gallant, A.R. (1981). On the bias in flexible functional forms and an essentially unbiased form: The Fourier flexible form, *Journal of Econometrics*, **15**(2): 211–245.

Gao, J. (2007). *Nonlinear Time Series: Semiparametric and Nonparametric Methods*, Chapman & Hall, London.

Gao, J. and Hawthorne, K. (2006). Semiparametric estimation and testing of the trend of temperature series, *The Econometrics Journal*, **9**(2): 332–355.

Gao, J. and Phillips, P.C.B. (2013). Semiparametric estimation in triangular system equations with nonstationarity, *Journal of Econometrics*, **176**(1): 59–79.

Gao, J. and Robinson, P.M. (2016). Inference on nonstationary time series with moving mean, *Econometric Theory*, **32**(2): 431–457.

Gao, J., Linton, O. and Peng, B. (2020). Inference on a semiparametric model with global power law and local nonparametric trends, *Econometric Theory*, **36**(2): 223–249.

Gao, J., King, M.L., Lu, Z. and Tjøstheim, D. (2009). Specification testing in nonlinear and nonstationary time series autoregression, *The Annals of Statistics*, **37**(6B): 3893–3928.

Gay, C., Estrada, F. and Sánchez, A. (2009). Global and hemispheric temperatures revisited, *Climatic Change*, **94**(3–4): 333–349.

Gettelman, A. and Rood, R.B. (2016). *Demystifying Climate Models: A Users Guide to Earth System Models*, Springer, Berlin.

Gil-Alana, L.A. (2003). Testing of fractional cointegration in macroeconomic time series, *Oxford Bulletin of Economics and Statistics*, **65**(4): 517–529.

Gil-Alana, L.A. (2008). Time trend estimation with breaks in temperature time series, *Climatic Change*, **89**(3–4): 325–337.

Gil-Alana, L.A. and Hualde, J. (2009). Fractional integration and cointegration: An overview and an empirical application. In *Palgrave Handbook of Econometrics: Volume 2: Applied Econometrics*, Palgrave Macmillan, New York, New York, pp. 434–469.

Giraitis, L., Kapetanios, G., and Yates, T. (2014). Inference on stochastic time-varying coefficient models, *Journal of Econometrics*, **179**(1): 46–65.

Gomez, V. (2001). The use of butterworth filters for trend and cycle estimation in economic time series, *Journal of Business and Economic Statistics*, **19**(3): 365–373.

Granger, C.W.J. (1980). Long memory relationships and the aggregation of dynamic models, *Journal of Econometrics*, **14**(2): 227–238.

Granger, C.W.J. (1981). Some properties of time series data and their use in econometric model specification, *Journal of Econometrics*, **16**(1): 121–130.

Granger, C.W.J. (1983). Co-integrated variables and error-correcting models, Discussion Paper 83–13, Department of Economics, University of California, San Diego.

Granger, C.W.J. and Joyeux, R. (1980). An introduction to long-memory time series models and fractional differencing, *Journal of Time Series Analysis*, **1**(1): 15–29.

Granger, C.W.J. and Newbold, P. (1974). Spurious regressions in econometrics, *Journal of Econometrics*, **2**(2): 111–120.

Granger, C.W.J. and Weiss, A.A. (1983). Time series analysis of error-correction models. In *Studies in Econometrics, Time Series, and Multivariate Statistics*, Academic Press, New York, pp. 255–278.

Gregoir, S. (2010). Fully modified estimation of seasonally cointegrated processes, *Econometric Theory*, **26**(5): 1491–1528.

Gregory, A.W. and Hansen, B.E. (1996). Practitioners corner: Tests for cointegration in models with regime and trend shifts, *Oxford Bulletin of Economics and Statistics*, **58**(3): 555–560.

Grenander, U. and Rosenblatt, M. (1954). Regression analysis of time series with stationary residuals, *Proceedings of the National Academy of Social Sciences*, **40**(9): 812–816.

Grenander, U. and Rosenblatt, M. (2008). *Statistical Analysis of Stationary Time Series*, Vol. 320, American Mathematical Society, New York.

Hall, R.E. (1978). Stochastic implications of the life cycle-permanent income hypothesis: Theory and evidence, *The Journal of Political Economy*, **86**(6): 971–987.

Hallin, M., La Vecchia, D., and Liu, H. (2022). Center-outward r-estimation for semiparametric varma models, *Journal of the American Statistical Association*, **117**(538): 925–938.

Hamilton, J.D. (1994). *Time Series Analysis*, Princeton University Press, New Jersey.

Hamilton, J.D. (2018). Why you should never use the Hodrick-Prescott filter, *The Review of Economics and Statistics*, **100**(5): 831–843.

Han, H. and Ogaki, M. (1997). Consumption, income and cointegration, *International Review of Economics and Finance*, **6**(2): 107–117.

Hannachi, A. (2007). Pattern hunting in climate: A new method for finding trends in gridded climate data, *International Journal of Climatology*, **27**(1): 1–15.

Hansen, B.E. (1992). Tests for parameter instability in regressions with I(1) processes, *Journal of Business and Economic Statistics*, **10**(3): 321–335.

Hansen, B.E. and Phillips, P.C.B. (1990). Estimation and inference in models of cointegration: A simulation study, *Advances in Econometrics*, **8**: 225–248.

Härdle, W., Liang, H. and Gao, J. (2000). *Partially Linear Models*, Physica-Verlag, Heidelberg.

Härdle, W., Müller, M., Sperlich, S. and Werwatz, A. (2004). *Nonparametric and Semiparametric Models*, Vol. 1, Springer, Berlin.

Härdle, W. and Vieu, P. (1992). Kernel regression smoothing of time series, *Journal of Time Series Analysis*, **13**(3): 209–232.

Hargreaves, C. (1994). *Non-stationary Time Series Analysis and Cointegration*, Oxford University Press, Oxford.

Harris, D., Harvey, D.I., Leybourne, S.J. and Taylor, A.R. (2009). Testing for a unit root in the presence of a possible break in trend, *Econometric Theory*, **25**(6): 1545–1588.

Harvey, A. (1997). Trends, cycles and autoregressions, *The Economic Journal*, **107**(440): 192–201.

Harvey, A.C. (1990). *Forecasting, Structural Time Series Models and the Kalman Filter*, Cambridge University Press, Cambridge.

Harvey, A.C. and Phillips, G. (1982). The estimation of regression models with time-varying parameters. In *Games, Economic Dynamics, and Time Series Analysis: A Symposium in Memoriam Oskar Morgenstern Organized at the Institute for Advanced Studies, Vienna*, Physica, Heidelberg, pp. 306–321.

Harvey, D.I., Leybourne, S.J. and Taylor, A.R. (2014). Unit root testing under a local break in trend using partial information on the break date, *Oxford Bulletin of Economics and Statistics*, **76**(1): 93–111.

Harvey, D.I., Leybourne, S.J. and Taylor, A.R. (2007). A simple, robust and powerful test of the trend hypothesis, *Journal of Econometrics*, **141**(2): 1302–1330.

Harvey, D.I., Leybourne, S.J. and Taylor, A.R. (2009). Simple, robust, and powerful tests of the breaking trend hypothesis, *Econometric Theory*, **25**(4): 995–1029.

Harvey, D.I., Leybourne, S.J., and Taylor, A.R. (2012). Unit root testing under a local break in trend, *Journal of Econometrics*, **167**(1): 140–167.

Hastie, T. and Tibshirani, R. (2018). Varying-Coefficient Models, *Journal of the Royal Statistical Society: Series B*, **55**(4): 757–779.

Hatemi-J, A. (2008). Tests for cointegration with two unknown regime shifts with an application to financial market integration, *Empirical Economics*, **35**(3): 497–505.

Hendry, D.F. (2000). *Econometrics: Alchemy or Science?: Essays in Econometric Methodology*, Oxford University Press, Oxford.

Hodrick, R.J. and Prescott, E.C. (1997). Postwar U.S. business cycles: An empirical investigation, *Journal of Money, Credit, and Banking*, **29**(1): 1–16.

Hong, S.H. and Phillips, P.C.B. (2010). Testing linearity in cointegrating relations with an application to purchasing power parity, *Journal of Business and Economic Statistics*, **28**(1): 96–114.

Horowitz, J.L. (2012). *Semiparametric Methods in Econometrics*, Vol. 131, Springer, New York.

Hsiang, S.M. (2016). Climate econometrics, *Annual Review of Resource Economics*, **8**: 43–75.

Hsiang, S.M., Burke, M. and Miguel, E. (2013). Quantifying the influence of climate on human conflict, *Science*, **341**(6151): 1235367.

Hsiao, C. and Li, Q. (2001). A consistent test for conditional heteroskedasticity in time-series regression models, *Econometric Theory*, **17**(1): 188–221.

Hu, Z., Phillips, P.C.B. and Wang, Q. (2021). Nonlinear cointegrating power function regression with endogeneity, *Econometric Theory*, **37**(6): 1173–1213.

Hualde, J. (2006). Unbalanced cointegration, *Econometric Theory*, **22**(5): 765–814.

Hualde, J. (2014). Estimation of long-run parameters in unbalanced cointegration, *Journal of Econometrics*, **178**(2): 761–778.

Hualde, J. and Robinson, P.M. (2010). Semiparametric inference in multivariate fractionally cointegrated systems, *Journal of Econometrics*, **157**(2): 492–511.

Hylleberg, S., Engle, R.F., Granger, C.W.J. and Yoo, B.S. (1990). Seasonal integration and cointegration, *Journal of Econometrics*, **44**(1–2): 215–238.

Hyndman, R.J., King, M.L. Pitrun, I. and Billah, B. (2005). Local linear forecasts using cubic smoothing splines, *Australian and New Zealand Journal of Statistics*, **47**(1): 87–99.

Hyndman, R.J., Koehler, A.B., Ord, J.K. and Snyder, R.D. (2008). *Forecasting with Exponential Smoothing: The State Space Approach*, Springer, Berlin.

IPCC (2014). Summary for policymakers. In *Climate Change 2014: Synthesis Report*, IPCC, Geneva, pp. 1–34.

Jenish, N. (2016). Spatial semiparametric model with endogenous regressors, *Econometric Theory*, **32**(3): 714–739.

Jensen, M.J. and Maheu, J.M. (2013). Bayesian semiparametric multivariate garch modeling, *Journal of Econometrics*, **176**(1): 3–17.

Johansen, S. (1988). Statistical analysis of cointegration vectors, *Journal of Economic Dynamics and Control*, **12**(2–3): 231–254.

Johansen, S. (1991). Estimation and hypothesis testing of cointegration vectors in Gaussian vector autoregressive models, *Econometrica*, **59**(6): 1551–1580.

Johansen, S. (1995). *Likelihood-based Inference in Cointegrated Vector Autoregressive Models*, Oxford University Press, Oxford.

Johansen, S. and Juselius, K. (1990). Maximum likelihood estimation and inference on cointegration—with applications to the demand for money, *Oxford Bulletin of Economics and Statistics*, **52**(2): 169–210.

Johansen, S. and Juselius, K. (1992). Testing structural hypotheses in a multivariate cointegration analysis of the PPP and the UIP for UK, *Journal of Econometrics*, **53**: 1–3.

Johansen, S. and Nielsen, M.Ø. (2012). Likelihood inference for a fractionally cointegrated vector autoregressive model, *Econometrica*, **80**(6): 2667–2732.

Johansen, S. and Nielsen, M.Ø. (2018). Testing the CVAR in the fractional CVAR model, *Journal of Time Series Analysis*, **39**(6): 836–849.

Johansen, S. and Nielsen, M.Ø. (2019). Nonstationary cointegration in the fractionally cointegrated VAR model, *Journal of Time Series Analysis*, **40**(4): 519–543.

Johansen, S. and Schaumburg, E. (1999). Likelihood analysis of seasonal cointegration, *Journal of Econometrics*, **88**(2): 301–339.

Juhl, T. and Xiao, Z. (2005). A nonparametric test for changing trends, *Journal of Econometrics*, **127**(2): 179–199.

Kaufmann, R.K., Kauppi, H., Mann, M.L. and Stock, J.H. (2013). Does temperature contain a stochastic trend: Linking statistical results to physical mechanisms, *Climatic Change*, **118**: 729–743.

Kaufmann, R.K., Kauppi, H. and Stock, J.H. (2006). Emissions, concentrations, & temperature: A time series analysis, *Climatic Change*, **77**: 249–278.

Kaufmann, R.K., Kauppi, H. and Stock, J.H. (2010). Does temperature contain a stochastic trend? Evaluating conflicting statistical results, *Climatic Change*, **101**: 395–405.

Kaufmann, R.K. and Stern, D.I. (2002). Cointegration analysis of hemispheric temperature relations, *Journal of Geophysical Research*, **107**(D2): ACL 8–1 to 8–10.

Kim, C.-J. and Nelson, C.R. (1999). *State-Space Models with Regime Switching: Classical and Gibbs-Sampling Approaches with Applications, MIT Press Books*, Vol. 1, The MIT Press, Massachusetts.

Kim, C.-J. and Nelson, C.R. (2006). Estimation of a forward-looking monetary policy rule: A time-varying parameter model using ex post data, *Journal of Monetary Economics*, **53**(8): 1949–1966.

Kim, C.S. and Kim, I.M. (2012). Partial parametric estimation for nonstationary nonlinear regressions, *Journal of Econometrics*, **167**(2): 448–457.

Kim, D., Oka, T., Estrada, F. and Perron, P. (2020). Inference related to common breaks in a multivariate system with joined segmented trends with applications to global and hemispheric temperatures, *Journal of Econometrics*, **214**(1): 130–152.

King, R.G., Plosser, C.I., Stock, J.H. and Watson, M.W. (1991). Stochastic trends and economic fluctuations, *The American Economic Review*, **81**(4): 819–840.

Knudsen, M.F., Seidenkrantz, M.-S., Jacobsen, B.H., and Kuijpers, A. (2011). Tracking the Atlantic Multidecadal Oscillation through the last 8,000 years, *Nature Communications*, **2**: 178.

Kwiatkowski, D., Phillips, P.C.B., Schmidt, P. and Shin, Y. (1992). Testing the null hypothesis of stationarity against the alternative of a unit root: How sure are we that economic time series have a unit root? *Journal of Econometrics*, **54**(1): 159–178.

Lanzante, J.R. (1996). Resistant, robust and non-parametric techniques for the analysis of climate data: Theory and examples, including applications to historical radiosonde station data, *International Journal of Climatology*, **16**(11): 1197–1226.

Lavergne, P. and Vuong, Q. (2000). Nonparametric significance testing, *Econometric Theory*, **16**(4): 576–601.

Lee, H.S. (1992). Maximum likelihood inference on cointegration and seasonal cointegration, *Journal of Econometrics*, **54**(1–3): 1–47.

Lettau, M. and Ludvigson, S. (2001). Consumption, aggregate wealth, and expected stock returns, *The Journal of Finance*, **56**(3): 815–849.

Lettau, M. and Van Nieuwerburgh, S. (2008). Reconciling the return predictability evidence, *The Review of Financial Studies*, **21**(4): 1607–1652.

Li, D., Chen, J. and Gao, J. (2011). Non-parametric time-varying coefficient panel data models with fixed effects, *The Econometrics Journal*, **14**(3): 387–408.

Li, D., Linton, O., and Lu, Z. (2015). A flexible semiparametric forecasting model for time series, *Journal of Econometrics*, **187**(1): 345–357.

Li, D., Phillips, P.C.B., and Gao, J. (2020). Kernel-based inference in time-varying coefficient cointegrating regression, *Journal of Econometrics*, **215**(2): 607–632.

Li, D., Tjøstheim, D. and Gao, J. (2016). Estimation in nonlinear regression with Harris recurrent Markov chains, *The Annals of Statistics*, **44**(5): 1957–1987.

Li, Q. and Racine, J.S. (2007). *Nonparametric Econometrics: Theory and Practice*, Princeton University Press, New Jersey.

Lin, Y. and Tu, Y. (2020). Robust inference for spurious regressions and cointegrations involving processes moderately deviated from a unit root, *Journal of Econometrics*, **219**(1): 52–65.

Lin, Y., Tu, Y. and Yao, Q. (2020). Estimation for double-nonlinear cointegration, *Journal of Econometrics*, **216**(1): 175–191.

Lineesh, M. and John, C.J. (2010). Analysis of non-stationary time series using wavelet decomposition, *Nature and Science*, **8**(1): 53–59.

Magnus, J.R., Melenberg, B., and Muris, C. (2011). Global warming and local dimming: The statistical evidence, *Journal of the American Statistical Association*, **106**(494): 452–464.

Maki, D. (2012). Tests for cointegration allowing for an unknown number of breaks, *Economic Modelling*, **29**(5): 2011–2015.

Mankiw, N.G. and Shapiro, M.D. (1985). Trends, random walks, and tests of the permanent income hypothesis, *Journal of Monetary Economics*, **16**(2): 165–174.

Mann, M.E., Steinman, B.A., Brouillette, D.J. and Miller, S.K. (2021). Multidecadal climate oscillations during the past millennium driven by volcanic forcing, *Science*, **371**(6533): 1014–1019.

McGuffie, K. and Henderson-Sellers, A. (2014). *The Climate Modelling Primer*, John Wiley & Sons, Blackwell.

McKitrick, R., McIntyre, S., and Herman, C. (2010). Panel and multivariate methods for tests of trend equivalence in climate data series, *Atmospheric Science Letters*, **11**(4): 270–277.

Mills, T.C. (2009). How robust is the long-run relationship between temperature and radiative forcing? *Climatic Change*, **94**(3–4): 351–361.

Mills, T.C. (2019). *Applied Time Series Analysis: A Practical Guide to Modeling and Forecasting*, Academic Press, London.

Mitchell, J. and Karoly, D. (2001). Detection of climate change and attribution of causes. In *Climate Change 2001: The Scientific Basis*, Cambridge University Press, Cambridge, 427–431.

Morton, R., Kang, E.L., and Henderson, B.L. (2009). Smoothing splines for trend estimation and prediction in time series, *Environmetrics*, **20**(3): 249–259.

Mudelsee, M. (2019). Trend analysis of climate time series: A review of methods, *Earth-Science Reviews*, **190**: 310–322.

Müller, U.K. (2008). The impossibility of consistent discrimination between I(0) and I(1) processes, *Econometric Theory*, **24**(3): 616–630.

Müller, U.K. and Watson, M.W. (2018). Long-run covariability, *Econometrica*, **86**(3): 775–804.

Nelson, C.R. and Kang, H. (1981). Spurious periodicity in inappropriately detrended time series, *Econometrica*, **49**(3): 741–751.

Nelson, C.R. and Kang, H. (1984). Pitfalls in the use of time as an explanatory variable in regression, *Journal of Business and Economic Statistics*, **2**(1): 73–82.

Nelson, C.R. and Plosser, C.R. (1982). Trends and random walks in macroeconmic time series: Some evidence and implications, *Journal of Monetary Economics*, **10**(2): 139–162.

Neuman, C.P. and Schonbach, D.I. (1974). Discrete (Legendre) orthogonal polynomials — a survey, *International Journal for Numerical Methods in Engineering*, **8**(4): 743–770.

Newey, W.K. and West, K.D. (1987). A simple, positive semi-definite, heteroskedasticity and autocorrelation consistent covariance matrix, *Econometrica*, **55**(3): 703–708.

Newey, W.K. and West, K.D. (1994). Automatic lag selection in covariance matrix estimation, *The Review of Economic Studies*, **61**(4): 631–653.

Nicholls, D.F. and Pagan, A. (1985). Varying coefficient regression, *Handbook of Statistics*, **5**: 413–449.

Nordhaus, W.D. (2013). *The Climate Casino: Risk, Uncertainty, and Economics for a Warming World*, Yale University Press, New Haven.

Ouliaris, S., Park, J.Y. and Phillips, P.C.B. (1989). Testing for a unit root in the presence of a maintained trend. In *Advances in Econometrics and Modelling*, Springer, London, pp. 7–28.

Park, J.Y. and Hahn, S.B. (1999). Cointegrating regressions with time varying coefficients, *Econometric Theory*, **15**(5): 664–703.

Park, J.Y. and Phillips, P.C.B. (1988). Statistical inference in regressions with integrated processes: Part 1, *Econometric Theory*, **4**(3): 468–497.

Park, J.Y. and Phillips, P.C.B. (1999). Asymptotics for nonlinear transformations of integrated time series, *Econometric Theory*, **15**(3): 269–298.

Park, J.Y. and Phillips, P.C.B. (2001). Nonlinear regressions with integrated time series, *Econometrica*, **69**(1): 117–161.

Patton, A.J., Ziegel, J.F. and Chen, R. (2019). Dynamic semiparametric models for expected shortfall (and value-at-risk), *Journal of Econometrics*, **211**(2): 388–413.

Perron, P. (1988). Trends and random walks in macroeconomic time series: Further evidence from a new approach, *Journal of Economic Dynamics and Control*, **12**(2–3): 297–332.

Perron, P. (1989). The great crash, the oil price shock, and the unit root hypothesis, *Econometrica*, **57**(7): 1361–1401.

Perron, P. and Yabu, T. (2009). Estimating deterministic trends with an integrated or stationary noise component, *Journal of Econometrics*, **151**(1): 56–69.

Pfaff, B. (2008). *Analysis of Integrated and Cointegrated Time Series with R*, Springer, New York.

Phillips, P.C.B. (1986). Understanding spurious regressions in econometrics, *Journal of Econometrics*, **33**(3): 311–340.

Phillips, P.C.B. (1991). Optimal inference in cointegrated systems, *Econometrica*, **59**(2): 283–306.

Phillips, P.C.B. (1998). New tools for understanding spurious regressions, *Econometrica*, **66**(6): 1299–1325.

Phillips, P.C.B. (2001). Trending time series and macroeconomic activity: Some present and future challenges, *Journal of Econometrics*, **100**(1): 21–27.

Phillips, P.C.B. (2003). Laws and limits of econometrics, *The Economic Journal*, **113**(486): C26–C52.

Phillips, P.C.B. (2005). Challenges of trending time series econometrics, *Mathematics and Computers in Simulation*, **68**(5): 401–416.

Phillips, P.C.B. (2007). Regression with slowly varying regressors and nonlinear trends, *Econometric Theory*, **23**(4): 557–614.

Phillips, P.C.B. (2009). Local limit theory and spurious nonparametric regression, *Econometric Theory*, **25**(6): 1466–1497.

Phillips, P.C.B. (2010). The mysteries of trend, *Macroeconomic Review*, **9**(2): 82–89.

Phillips, P.C.B. and Hansen, B.E. (1990). Statistical inference in instrumental variables regression with I(1) processes, *The Review of Economic Studies*, **57**(1): 99–125.

Phillips, P.C.B. and Jin, S. (2021). Business cycles, trend elimination, and the HP filter, *International Economic Review*, **62**(2): 469–520.

Phillips, P.C.B. and Ouliaris, S. (1988). Testing for cointegration using principal components methods, *Journal of Economic Dynamics and Control*, **12**(2–3): 205–230.

Phillips, P.C.B. and Ouliaris, S. (1990). Asymptotic properties of residual based tests for cointegration, *Econometrica*, **58**(1): 165–193.

Phillips, P.C.B. and Perron, P. (1988). Testing for a unit root in time series regression, *Biometrika*, **75**(2): 335–346.

Phillips, P.C.B. and Wang, Y. (2021). When bias contributes to variance: True limit theory in functional coefficient cointegrating regression, *Journal of Econometrics*, **232**(2): 469–489.

Phillips, P.C.B. and Xiao, Z. (1998). A primer on unit root testing, *Journal of Economic Surveys*, **12**(5): 423–470.

Phillips, P.C.B., Li, D. and Gao, J. (2017). Estimating smooth structural change in cointegration models, *Journal of Econometrics*, **196**(1): 180–195.

Pretis, F. (2020). Econometric modelling of climate systems: The equivalence of energy balance models and cointegrated vector autoregressions, *Journal of Econometrics*, **214**(1): 256–272.

Pretis, F. (2021). Exogeneity in climate econometrics, *Energy Economics*, **96**: 105122.

Pretis, F. and Hendry, D.F. (2013). Some hazards in econometric modelling of climate change, *Earth System Dynamics*, **4**(2): 375–384.

Priestley, M.B. (1988). *Non-linear and Non-stationary Time Series Analysis* (Academic Press, London, London).

Quintos, C.E. and Phillips, P.C.B. (1993). Parameter constancy in cointegrating regressions, *Empirical Economics*, **18**(4): 675–706.

Rao, B.B. (2010). Deterministic and stochastic trends in the time series models: A guide for the applied economist, *Applied Economics*, **42**(17): 2193–2202.

Ren, Y., Tu, Y., and Yi, Y. (2019). Balanced predictive regressions, *Journal of Empirical Finance*, **54**: 118–142.

Robinson, P.M. (1988). Root-N-consistent semiparametric regression, *Econometrica*, **56**(4): 931–954.

Robinson, P.M. (1989). Nonparametric estimation of time-varying parameters. In *Statistical Analysis and Forecasting of Economic Structural Change*, Springer, Berlin, pp. 253–264.

Robinson, P.M. (1991). Time-varying nonlinear regression. In *Economic Structural Change: Analysis and Forecasting*, Springer, Berlin, pp. 179–190.

Robinson, P.M. (1994). Efficient tests of nonstationary hypotheses, *Journal of the American Statistical Association*, **89**(428): 1420–1437.

Robinson, P.M. (1997). Large-sample inference for nonparametric regression with dependent errors, *The Annals of Statistics*, **25**(5): 2054–2083.

Robinson, P.M. (2012a). Inference on power law spatial trends, *Bernoulli*, **18**(2): 644–677.

Robinson, P.M. (2012b). Nonparametric trending regression with cross-sectional dependence, *Journal of Econometrics*, **169**(1): 4–14.

Rosenberg, B. (1973). Random coefficients models: The analysis of a cross section of time series by stochastically convergent parameter regression. In *Annals of Economic and Social Measurement*, Volume 2, Number 4 (NBER), pp. 399–428.

Rudebusch, G.D. (1993). The uncertain unit root in real GNP, *The American Economic Review*, **83**(1): 264–272.

Saart, P., Gao, J. and Kim, N.H. (2014). Semiparametric methods in nonlinear time series analysis: A selective review, *Journal of Nonparametric Statistics*, **26**(1): 141–169.

Said, S.E. and Dickey, D.A. (1984). Testing for unit roots in autoregressive-moving average models of unknown order, *Biometrika*, **71**(3): 599–607.

Sakarya, N. and De Jong, R.M. (2020). A property of the Hodrick–Prescott filter and its application, *Econometric Theory*, **36**(5): 840–870.

Sargan, J.D. (1958). The estimation of economic relationships using instrumental variables, *Econometrica*, **26**(3): 393–415.

Schwert, G.W. (1989). Tests for unit roots: A Monte Carlo investigation, *Journal of Business and Economic Statistics*, **7**(2): 147–159.

Shimotsu, K. and Phillips, P.C.B. (2005). Exact local Whittle estimation of fractional integration, *The Annals of Statistics*, **33**(4): 1890–1933.

Skog, O.J. (1988). Testing causal hypotheses about correlated trends: Pitfalls and remedies, *Contemporary Drug Problems*, **15**(4): 565–606.

Stambaugh, R.F. (1999). Predictive regressions, *Journal of Financial Economics*, **54**(3): 375–421.

Stein, M.L. (2020). Some statistical issues in climate science, *Statistical Science*, **35**: 31–41.

Stern, D.I. and Kaufmann, R.K. (1999). Econometric analysis of global climate change, *Environmental Modelling and Software*, **14**(6): 597–605.

Stern, N. (2016). Economics: Current climate models are grossly misleading, *Nature*, **530**(7591): 407–409.

Stock, J.H. (1987). Asymptotic properties of least squares estimators of cointegrating vectors, *Econometrica*, **55**(5): 1035–1056.

Stock, J.H. (1994). Deciding between I(1) and I(0), *Journal of Econometrics*, **63**(1): 105–131.

Stock, J.H. and Watson, M.W. (1988). Testing for common trends, *Journal of the American Statistical Association*, **83**(404): 1097–1107.

Stock, J.H. and Watson, M.W. (1996). Evidence on structural instability in macroeconomic time series relations, *Journal of Business and Economic Statistics*, **14**(1): 11–30.

Storelvmo, T., Leirvik, T., Lohmann, U., Phillips, P.C.B. and Wild, M. (2016). Disentangling greenhouse warming and aerosol cooling to reveal earth's climate sensitivity, *Nature Geoscience*, **9**: 286–289.

Su, L. and Ullah, A. (2006). Profile likelihood estimation of partially linear panel data models with fixed effects, *Economics Letters*, **92**(1): 75–81.

Sun, Y. (2011). Robust trend inference with series variance estimator and testing-optimal smoothing parameter, *Journal of Econometrics*, **164**(2): 345–366.

Sun, Y., Cai, Z. and Li, Q. (2013). Semiparametric functional coefficient models with integrated covariates, *Econometric Theory*, **29**(3): 659–672.

Sun, Y., Cai, Z. and Li, Q. (2016). A consistent nonparametric test on semiparametric smooth coefficient models with integrated time series, *Econometric Theory*, **32**(4): 988–1022.

Sun, Y., Carroll, R.J. and Li, D. (2009). Semiparametric estimation of fixed-effects panel data varying coefficient models. In *Nonparametric Econometric Methods*, Emerald Group Publishing Limited, Leeds, pp. 101–129.

Swamy, P.A. and Tinsley, P.A. (1980). Linear prediction and estimation methods for regression models with stationary stochastic coefficients, *Journal of Econometrics*, **12**(2): 103–142.

Teräsvirta, T., Tjøstheim, D. and Granger, C.W.J. (2010). *Modelling Nonlinear Economic Time Series*, Oxford University Press, Oxford.

Tjøstheim, D. (2020). Some notes on nonlinear cointegration: A partial review with some novel perspectives, *Econometric Reviews*, **39**(7): 655–673.

Tsay, R.S. and Chen, R. (2018). *Nonlinear Time Series Analysis*, Vol. 891, John Wiley & Sons, New Jersey.

Tu, Y. and Wang, Y. (2022). Spurious functional-coefficient regression models and robust inference with marginal integration, *Journal of Econometrics*, **229**(2): 396–421.

Vahid, F. and Engle, R.F. (1993). Common trends and common cycles, *Journal of Applied Econometrics*, **8**(4): 341–360.

Vogelsang, T.J. and Franses, P.H. (2005). Testing for common deterministic trend slopes, *Journal of Econometrics*, **126**(1): 1–24.

Wang, Q. and Phillips, P.C.B. (2009a). Asymptotic theory for local time density estimation and nonparametric cointegrating regression, *Econometric Theory*, **25**(3): 710–738.

Wang, Q. and Phillips, P.C.B. (2009b). Structural nonparametric cointegrating regression, *Econometrica*, **77**(6): 1901–1948.

Wang, Q. and Phillips, P.C.B. (2012). A specification test for nonlinear nonstationary models, *The Annals of Statistics*, **40**(2): 727–758.

Wang, Q. and Phillips, P.C.B. (2016). Nonparametric cointegrating regression with endogeneity and long memory, *Econometric Theory*, **32**(2): 359–401.

Wang, Q., Wu, D. and Zhu, K. (2018). Model checks for nonlinear cointegrating regression, *Journal of Econometrics*, **207**(2): 261–284.

Wang, T., Hamann, A., Spittlehouse, D.L. and Murdock, T.Q. (2012). ClimateWNA — high-resolution spatial climate data for western North America, *Journal of Applied Meteorology and Climatology*, **51**(1): 16–29.

Welch, I. and Goyal, A. (2008). A comprehensive look at the empirical performance of equity premium prediction, *The Review of Financial Studies*, **21**(4): 1455–1508.

White, H. and Granger, C.W.J. (2011). Consideration of trends in time series, *Journal of Time Series Econometrics*, **3**(1): article 2.

Woodward, W.A., Gray, H.L. and Elliott, A.C. (2017). *Applied Time Series Analysis with R*, CRC Press, Boca Raton, FL.

Xiao, Z. (2009). Functional-coefficient cointegration models, *Journal of Econometrics*, **152**(2): 81–92.

Xu, K.L. (2012). Robustifying multivariate trend tests to nonstationary volatility, *Journal of Econometrics*, **169**(2): 147–154.

Yatchew, A. (2003). *Semiparametric Regression for the Applied Econometrician*, Cambridge University Press, Cambridge.

Yu, D., Huang, D. and Chen, L. (2023). Stock return predictability and cyclical movements in valuation ratios, *Journal of Empirical Finance*, **72**: 36–53.

Yule, G.U. (1926). Why do we sometimes get nonsense-correlations between time-series? — A study in sampling and the nature of time-series, *Journal of the Royal Statistical Society*, **89**(1): 1–63.

Zhang, G.P. and Qi, M. (2005). Neural network forecasting for seasonal and trend time series, *European Journal of Operational Research*, **160**(2): 501–514.

Zhang, R., Robinson, P.M. and Yao, Q. (2019). Identifying cointegration by eigenanalysis, *Journal of the American Statistical Association*, **114**(526): 916–927.

Zhang, S., Okhrin, O., Zhou, Q.M. and Song, P.X.-K. (2016). Goodness-of-fit test for specification of semiparametric copula dependence models, *Journal of Econometrics*, **193**(1): 215–233.

Zhang, Y., Su, L. and Phillips, P.C.B. (2012). Testing for common trends in semi-parametric panel data models with fixed effects, *The Econometrics Journal*, **15**(1): 56–100.

Zheng, J.X. (1996). A consistent test of functional form via nonparametric estimation techniques, *Journal of Econometrics*, **75**(2): 263–289.

Zivot, E. and Andrews, D.W.K. (1992). Further evidence on the great crash, the oil-price shock, and the unit-root, *Journal of Business and Economic Statistics*, **10**(3): 251–270.

Author Index

N

O

P

Q

R

S

T

U

V

W

X

Y

Z

Subject Index

S

T

U

V

W

Printed in the United States
by Baker & Taylor Publisher Services